PHYSIOLOGY OF ANGIOSPERMS

PHYSIOLOGY OF ANGIOSPERMS

By
Dr. Pooja
Department of Botany
R.C.C. College
Ghaziabad (U.P.)

DISCOVERY PUBLISHING HOUSE
NEW DELHI-110002

First Published-2004

ISBN 81-7141-854-6

Published by

DISCOVERY PUBLISHING HOUSE
4831/24, Ansari Road, Prahlad Street,
Darya Ganj, New Delhi-110002 (India)
Phone: 23279245 • Fax: 91-11-23253475
E-mail:dphtemp@indiatimes.com

Printed at :
ARORA OFFSET PRESS
DELHI-110092

PREFACE

The present title is the unsightful compilation offering a comprehensive account of the fundamental principles of Physiology of Angiosperms. It incorporates and organizes the information concerning plant physiology from relevant and authentic sources. It is an humble attempt to present a connected and precise account of the subject matter of this important branch of Botany which forms an integral part of the studies undertaken by the undergraduate and postgraduate students of the subject. The principles have been outlined in considerable detail along with the fundamental facts and theories that explain the life processes of plants. It is a compilation work and embodies a fairly comprehensive treatment of the fundamental facts and aspects of morphology and anatomy. The purpose of this title is to provide students an authoritative and upto-date text in a very simple way, easy to grasp by those who do not have strong background of this subject. The present text provides a background of facts, terminology and internal structure of common plants around us and may safely be used as laboratory guide. Much emphasis has been given on the anatomical study of angiosperms. Ecological aspects have also been dealt with in sufficient details.

Most of illustrations have been taken from professional papers and standard books and journals. Subject matter has also been obtained from a large number of standard books, reviews and research papers by internationally renowned workers and established authors. That is why there can be no claim to originality except in the manner of treatment.

Though the author has taken special care to present a current account, yet he is fully aware of his limitation, and the readers may come across the mistakes of various types. For all types of mistakes he extend his due apology.

The author expresses his thanks to his friends and colleagues whose continuous inspirations have invited him to bring out this text.

The author expresses his gratitute to Mr. Wasan and staff of M/s Discovery Publishing House for their whole hearted co-operation in the publication of this book.

Constructive criticisms and suggestions for improvement of the book will be thankfully acknowledged.

Author

CONTENTS

1

FLOWER

The flower may be regarded as a leafy shoot highly specialized for the performance of reproductive functions. The function of a flower is to produce seed and fruit, and the various parts (stem and leaf organs) are specially suited to the performance of that function. The essential structures in the flower, therefore, are the organs, which are more immediately concerned in the production of seed. We cannot appreciate the significance or morphological nature of the organs without a knowledge of the reproductive organs of the Vascular Cryptograms and Gymnosperms, and some of the terms used in this chapter will only be understood when these plants have been considered. The axis (stem portion) of the flower usually shows two regions–the pedicel, and the receptacle. The pedicel is, popularly, the stalk of the flower. It may be present or absent. If present, the flower is *pedicellate*; if absent, *sessile*. The receptacle is the portion of the axis to which the floral leaves are attached.

In many flowers there are four sets or series of floral leaves. To the outside are the sepals; collectively, they constitute the calyx. Internal to these are the petals, constituting the corolla. Then come the stamens forming the androecium; and finally, in the centre of the flower, are the carpels, forming the gynaeceum or pistil. The buttercup (*Ranunculus*) flower will serve as a convenient type of introduction to these structures. In the buttercup the carpels are separate from each other, and each shows a hollow basal portion called the ovary, above which are the parts known as style and stigma. In many flowers the carpels are united and form a single ovary.

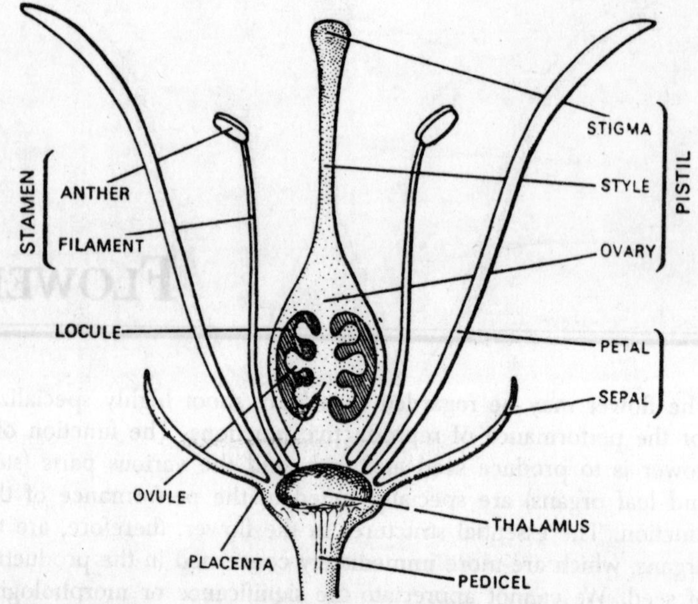

Fig. 1.1. A flower in L.S. showing position of different whorls on thalamus.

The following facts support the above view of the morphological character of the flower: (a) the flower, like an ordinary foliage-shoot, arises as a bud, very often in the axil of a leaf (*bract*). (b) The receptacle has the general structure of a stem, and the sepals and petals in their structure and development resemble leaves. (c) While in most cases the stamens and carpels are quite unlike leaves, there are certain conditions in which they become distinctly leaf-like. Thus, in many cultivated flowers, *e.g.* rose, the stamens are transformed into petals; in double cherry the gynaeceum is represented by a tuft of small green leaves; in water lily there is a gradual transition between petals and stamens.

Ontogeny of the Flower

The flower develops from an apical meristem that differs in only minor ways from that of a leafy stem. The early stages of floral-appendage development are closely like those of leaves. All organs develop laterally on the apical meristem, initiated by periclinal divisions below the surface layer. The sequence of development of the whorls of organs varies but it commonly acropetal. All organs have, at least for a time, an apical meristem, and those that are laterally expanded have marginal initials.

Procambium develops acropetally in all organs as does the phloem. The xylem matures either acropetally or both acropetally and basipetally from one or more points of beginning near the base of the organ. In carpels, especially in syncarpous ovaries, connection of the early-formed xylem with the xylem of the receptacle may be long delayed, even until the fruit is partly developed. The carpel remains an immature organ until the fruit is mature. In various parts it retains meristematic tissues of different types. The other organs are usually mature in the flower. Where fusion is present, the fusion may be ontogenetic, as it is frequently between carpel margins, or it may be phylogenetic, the fused organs arising from

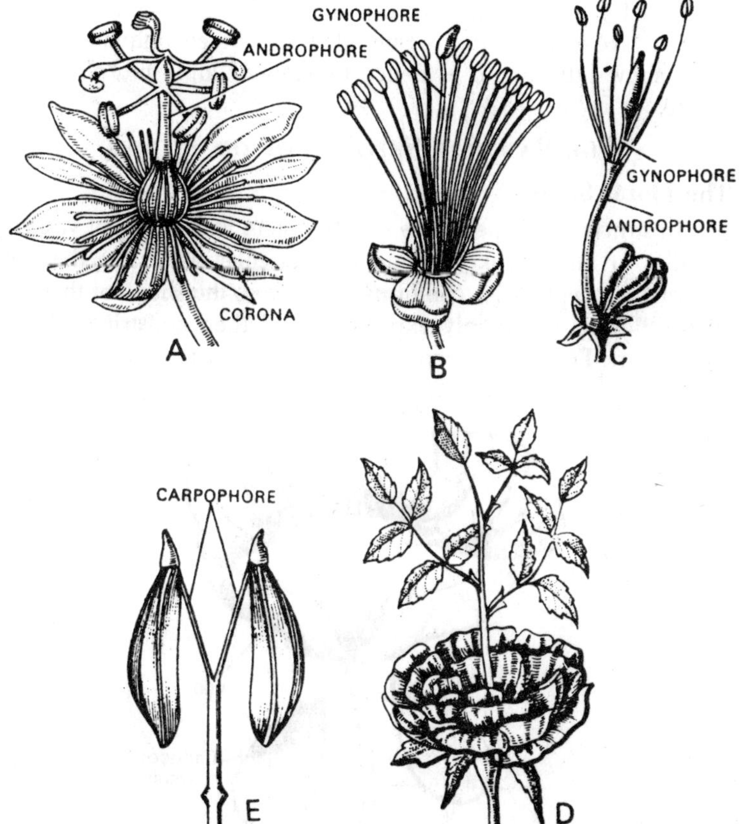

Fig. 1.2. A–Androphore of passion flower; B–Gynophore of Capparis; C–Androgynophore of Gynandropsis pentaphylla; D–Carpophore of fruit of Foeniculum; E–Rose showing monstrous development of thalamus.

a common meristem, as in typical inferior ovaries. The latter type of union has been called "congenital fusion." In adjacent organs that arise as independent organs from independent apical meristems, for example the members of a gamopetalous corolla, the initiating regions may shift from the apex to the base and merge, and the organs then continue growth as a single structure with only the vascular tissue showing its compound nature. Adnate organs, in any number, may arise independently but later continue growth from a single meristematic region as a fused but superficially single structure. Such apparently simple structures may show evidence of their compound nature in the presence of independent vascular bundles for every organ within the common structure, or the vascular skeletons of the organs may also be so intimately fused as to show little or no evidence of their multiple morphological nature.

THE VASCULAR SKELETON OF THE FLOWER

The Floral Stele

In structure, the pedicel is a typical stem with a ring of vascular bundles or an unbroken cylinder of vascular tissue. As the stele enters the receptacle, it conforms in shape to the shape of that part of the flower, commonly expanding and then constricting in its upper part. From the receptacular stele depart the vascular traces

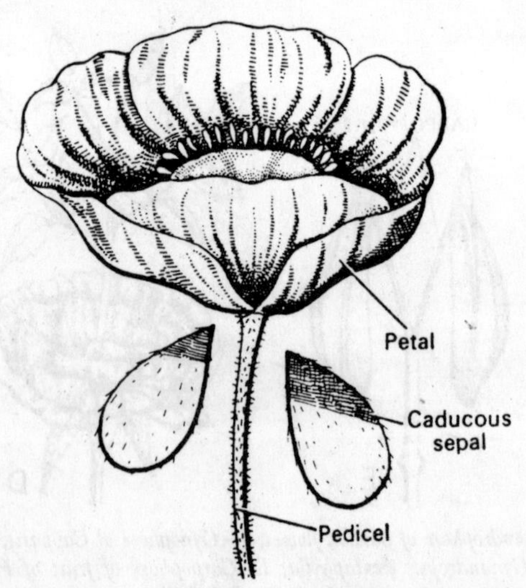

Fig. 1.3. Caducous sepals of Poppy.

to the various floral organs, traces which in origin, structure, and behaviour are similar to those of leaves. Gaps accompany the exit of these traces, and the gaps and the crowding of the organs break up the cylinder into a network of strands.

Often the vascular cylinder is further dissected, as in the stems of herbaceous plants with reduced vascular cylinders. The receptacular stele may be simple but is usually highly complex. The traces to sepals, petals, stamens, and carpels are given off successively from the stele in whorls or in spirals, according to the manner of arrangement of these organs in the flower. (Many organs that to external inspection are apparently whorled, such as the petals of some Ranunculaceae and the stamens of some Rosaceae, can be seen from anatomical evidence to be arranged in flat spirals.) The distal bundles of the receptacular stele commonly become the traces of the uppermost carpels, but in many floral types stelar bundles continue beyond the point where the last traces are given off and gradually fade out in the top of the receptacle. Such bundles are vestigial and usually consists of procambium or phloem only.

The Inflorescence

The floral or reproductive region of the plant is usually distinctly marked off from the foliage, or vegetative region, and is known as the inflorescence. Sometimes the main vegetative-axis of the plant ends in a single terminal flower, e.g. tulip and wood anemone. Here the flower is said to be solitary and terminal. In

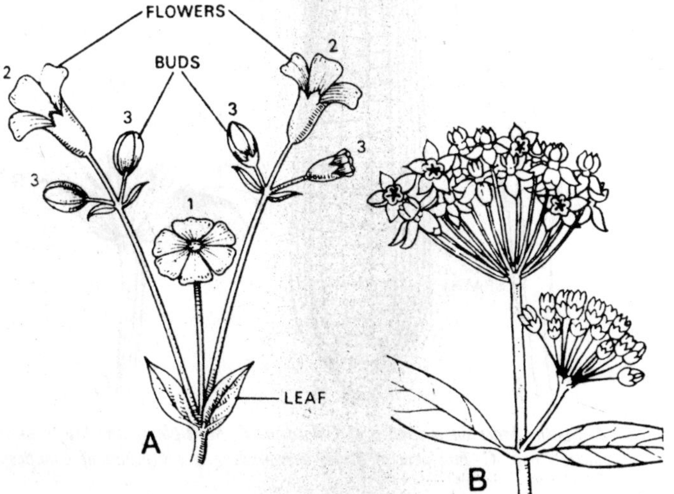

Fig. 1.4. A–Dichasial cyme Dianthus chinensis, B–Polychasial cyme Hamelia.

other cases, the flowers are developed singly in the axils of ordinary foliage leaves, and are called solitary and axillary. These are simple types of inflorescence. Usually the flowers are aggregated on a more or less complex branch-system.

According to the nature of the branching, and other features, many different kinds of such inflorescence are recognized. The main or primary axis of the inflorescence, together with any secondary axes which may be developed (apart from the pedicels of the flowers), is called the peduncle. This term is applied instead of pedicel to the stalks of solitary terminal, and solitary axillary flowers. If the peduncle is an unbranched leafless axis which arises from the midst of radical leaves and bears flowers at its apex, it is called a scape, e.g. cowslip and members of the Amaryllidaceae.

Bracts, etc.

When the flower arises as a lateral bud, the axis on which is borne is called the mother-axis. This may or may not be the primary axis of the inflorescence. The side of the flower which is towards the mother-axis (or towards the growing point of the mother-axis) is said to be posterior; the side away from the mother

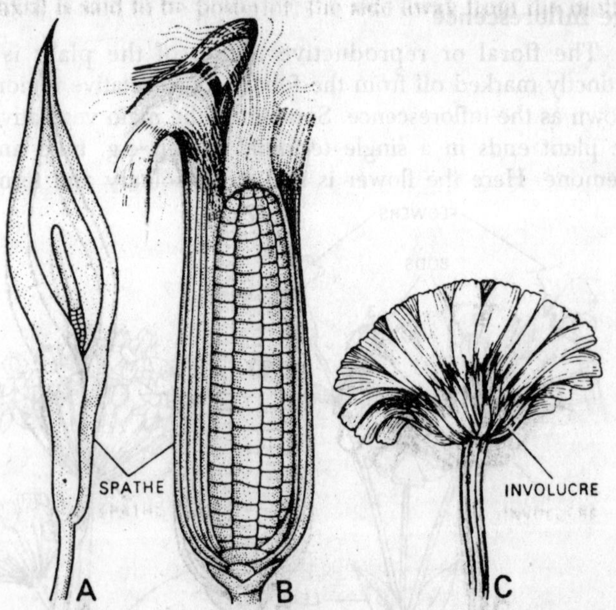

Fig. 1.5. A–Spathy bract on a spadix of Colocasia, B–Multiple spathy bracts covering a maize comb, C–Involucre of bracts surrounding a capitulum of Compositae.

axis is anterior. In a solitary terminal flower it is evident that these terms are not applicable. If the flower arises in the axil of a leaf-structure, this leaf structure is called a bract. Bracts present great variety of form and colour.

The bracts may be ordinary foliage leaves, as in solitary axillary flowers, or may more or less differ from them. Frequently they are small, green, and scale-like. In many plants they are reduced to small, tooth-like structures. When they are not green, but coloured like the petals of a flower, they are said to be *petaloid.* In many flowers the flower-stalk bears small outgrowths of the nature of reduced leaves. These are called bracteoles. When present, there are usually two in Dicotyledons, placed laterally, and one in Monocotyledons, situated on the posterior side.

Perianth

The outer series of floral leaves, distinct from stamens and carpels, constitute the perianth. Flowers without perianth are naked or achlamydeous. Flowers with a single perianth series are haplo- or monochlamydeous and the floral leaves (tepals) are either bracteoid or petaloid. Flowers with a double perianth (two series of floral leaves) are diplochlamydeous. When both series of floral leaves (tepals) are similar they are homoiochlamydeous, but fusion of the two series may render then apparently haplochlamydeous. When the two series of floral leaves are dissimilar, heterochlamydeous, the outer series is sepaloid (sepals, calyx) and the inner series petaloid (petals, corolla). When petals are absent as a result of abortion the flower is apopetalous.

The Essential Organs

The androecium the gynaeceum, because they bear the reproductive bodies, microspores (pollen-grains) and ovules, necessary for the production of seed, are called the essential organs. If both are present in the same flower (the rule in Angiosperms) the flower is hermaphrodite or *bisexual* (symbol ☿). When they are borne on different flowers, as is sometimes found, the flowers are imperfect or *unisexual.*

The unisexual flowers bearing the stamens are male (♂) or *staminate*; those bearing carpels, female (♀) or *pistillate.* If staminate and pistillate flowers are borne on the same plant (e.g. willow and some species of campion), dioecious. A plant is *polygamous* if it bears staminate, pistillate, and hermaphrodite flowers (e.g. *Fraxinus,*

Mangifera, *Anacardium*, *Rhus*, etc). Flowers in which both stamens and pistil have been lost are *neuter* (e.g. ray florets of *Senecio* species, *Volutarella*, *Helianthus*).

Floral Phyllotaxis

In most flowers the series of floral leaves are arranged in whorls, and the phyllotaxis is *cyclic*. Sometimes, however, all the floral leaves are in a spiral (e.g. cactus) and the flower is said to be *acyclic*. If some of the series are cyclically arranged, others spirally, the flowers are *hemicyclic*. In buttercup, for example, the calyx and corolla are whorled, while the stamens and carpels are spiral.

Number of Parts

In general, then, flowers have four definite series of floral leaves –calyx, corolla, androecium and gynaeceum–often with the same number of parts in each series. This, however, is far from true for many flowers where the number of parts in a series may be increased, or much reduced by the fusion of parts, or the loss of one or more parts from any series. The following examples will illustrate these points; *Viola* has five sepals, five petals, five stamens, three carpels; the pea has five sepals, five petals, ten stamens, one carpel; mustard has four sepals in two whorls, four petals in one whorl, six stamens in an outer whorl of two and an inner whorl of four, two sterile and two fertile carpels; many flowers have a large number of stamens in several whorls (e.g. *Rosa*).

It should be noticed that owning to the abbreviation of the receptacle and other causes, it is often difficult to distinguish the separate whorls, e.g. the two whorls of sepals in mustard, the two whorls of stamens in pea. Neglecting the reduction of parts met with in particular series, and more especially in the gynaeceum, we find that in Dicotyledons the series of floral leaves are, as a *rule*, arranged in twos, fours, or fives, or multiples of these numbers. In other words, the arrangement is *dimerous*, *tetramerous*, or *pentamerous*, rarely trimerous. The *trimerous* arrangement, i.e. in threes or multiples of three, is characteristic of Monocotyledons.

Alternation of Parts

The general rule is that the leaves of the different series alternate in position with each other–the petals alternate with the sepals, the stamens with the petals, etc. If there are several whorls of stamens, these whorls alternate with each other. But there are exceptions. In spiral flowers, the parts are sometimes superposed. In cyclic flowers the departure from regular alternation arises from various causes. In *Primula*, for example, there are five sepals, five petals, five stamens, and the stamens are opposite to the petals (*antipetalous*). Sometimes, where there are two alternating whorls of stamens, the outer whorl is opposite the petals. This is known as the *obdiplostemonous* condition. The carpels, where fewer in number, clearly cannot alternate with the leaves of the outer series.

Regular and Irregular Flowers

In regular flowers the parts in each series have the same size and form, *i.e.* the sepals resemble each other, so also the petals, etc. Irregular flowers are those in which some of the floral leaves in any one series have a different shape or size from the others– for example, the petals of pea or violet.

Floral Symmetry

Flowers may be radially symmetrical (actinomorphic), isobilateral (zygomorphic), or asymmetrical. The planes of symmetry may be median or antero-posterior, diagonal, or lateral. Zygomorphy is frequently due to irregularity, and this is the sense in which the term is used as a rule in descriptive botany. In zygomorphic flowers the plane of symmetry is, in most cases, antero-posterior or median, i.e. it is the plane passing through the anterior and posterior sides of the flower, e.g. pea, violet. Asymmetrical flowers are usually spiral, as for example in the Cactaceae.

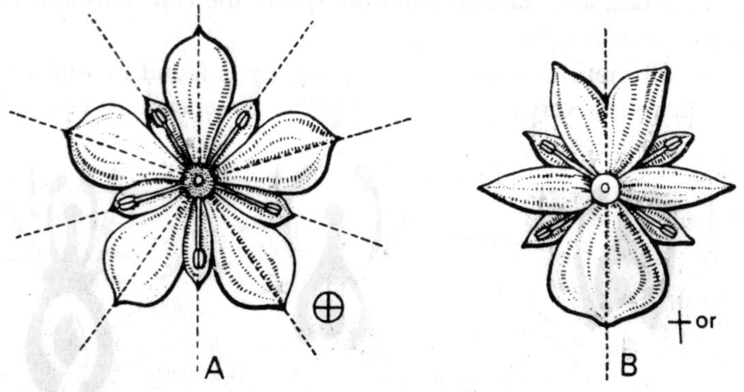

Fig. 1.6. Symmetry of flower. A–Actinomorphic, B–Zygomorphic.

The Receptacle : Insertion of Floral Leaves

The receptacle is nearly always short. Only occasionally is it elongated between the whorls of floral leaves, as in some species of campion. The form of the receptacle varies considerably. It may be convex and more or less dilated, or flattened, or hollow and cup-shaped. The insertion of floral leaves varies according to the form of the receptacle. In many flowers (e.g. *Ranunculus*, *Silene*, *Papaver*) the receptacle is more or less convex.

The gynaeceum is developed at the apex of the receptacle; the stamens, petals, and sepals, are inserted, in order, on the side of receptacle below the gynaeceum. This is to hypogynous arrangement. Suppose now that the receptacle is not convex, but forms a flattened circular disc. The apex of the receptacle is , of course, in the middle of the disc, and the flattened form is due to the sides of the receptacle having grown up to the same level. The gynaeceum is developed in the middle of the disc, and the sepals, petals, and stamens round the rim or margin. They are not *underneath* the gynaeceum, but round about it. Hence this is called a perigynous arrangement.

Sometimes the carpels are borne on a conical protuberance in the middle of the disc; this would represent a continued growth of the apex (e.g. *Fragaria*, *Rubus*, etc.). It is with the perigynous condition that the student will experience most difficulty; there are so many degrees of it. The receptacle may not be flat, but hollowed out, and more or less cup-like. This is due to the sides of the receptacle continuing to grow above the apex, which lies at the bottom of the cup. The carpels (gynaeceum) are developed in the cup; the sepals, petals, and stamens from the rim of the cup. This also is a perigynous condition.

A still more extreme form of perigyny is found in wild rose. Here there is a very deep cup. Finally, in the epigynous condition

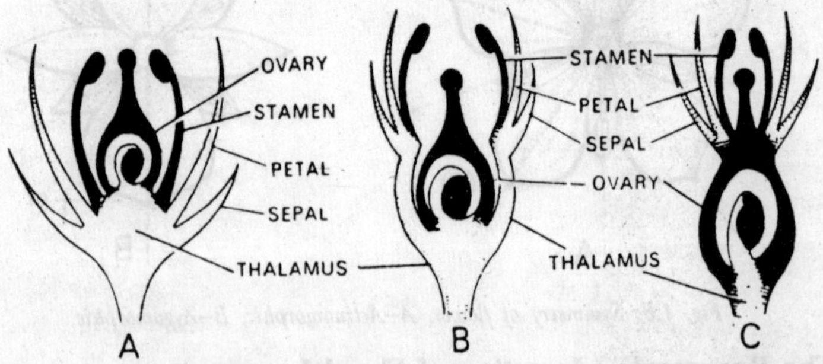

Fig. 1.7. Insertion of floral part of thalamus, A–Hypogynous (China rose), B–Perigynous (Rose), C–Epigynous (Strawberry).

the receptacle forms a deep cup as in the extreme forms of perigyny; but the carpels are from the first adherent to the receptacle, which is for this reason considered as part of the ovary. Thus in epigynous flowers the sepals, petals, and stamens are inserted on the gynaeceum. In the perigynous condition the receptacle wall remains distinct from the ovary.

Nectaries

The receptacle frequently bears a fleshy or glandular outgrowth, such as is found on the top of the inferior ovary in Umbelliferae and in the ivy (*Hedera*). This is termed the disc. In *Rubus* the disc lines the outer concave part of the receptacle. Usually the disc is lobed (vine), and frequently it secretes nectar. Nectaries may, however, develop from, or upon, any part of the flower. Thus in *Viola* the outgrowths borne by two stamens secrete nectar into a hollow spur borne by the anterior petal. In buttercup a small scale at the base of each of the petals covers a pocket-shaped nectary, while in christmas rose there are hollow tubular nectaries between the androecium and perianth. Nectaries occur upon the gynaeceum of *Gentiana* and each sepal of *Althaea* bears a nectary on its inner surface.

The Calyx

The calyx may consist of numerous sepals showing a primitive spiral arrangement, as in cactus and water lily; but usually it consists of from two to five sepals. If the sepals are free, the calyx is polysepalous. When they are united laterally, however slightly, the calyx is gamosepalous. The gamosepalous condition is due, not to the actual fusion of originally separate sepals, but to common basal growth during development. In all hypogynous and perigynous arrangements the calyx is described as *inferior*; in the epigynous flower the calyx is described as *superior*. In some flowers, e.g. *Fragaria* (strawberry), the sepals are stipulate. The stipules fuse in pairs between the sepals and produce an outer series of small sepal-like structures, forming what appears like an outer calyx. This is known as the epicalyx.

An epicalyx may also be produced by the aggregation of bracts or bracteoles beneath the calyx, e.g. Malvaceae, shoe flower. The calyx usually has a protective function. It commonly serves to protect the parts of the young flower in the bud. When the flower opens the calyx may fall off, e.g. in poppy, in which case it is said to be *caducous*, or the sepals fold back as in the wild rose. The

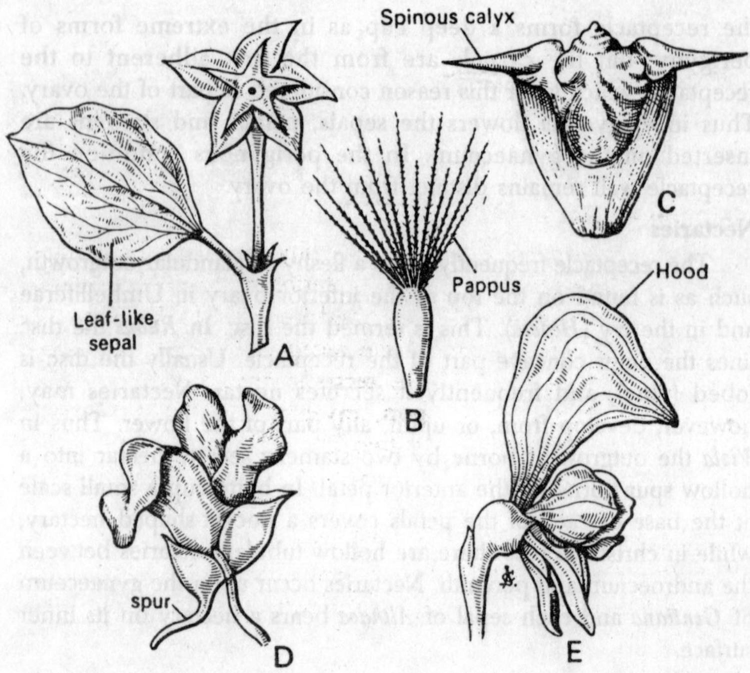

Fig. 1.8. Different forms of Calyx. A–Mussaenda–one sepal modified into a leaf like structure, B–Pappus in Mikania, C–Trapa bispinosa–spinous calyx, D–Impatiens–Spurred calyx, E–Aconitum–hooded calyx.

calyx is *deciduous* if it falls off when the flower withers. But frequently it persists till fruiting takes place and protects the young fruit, which is developed from the ovary of the flower (e.g. strawberry and Labiatae). A gamosepalous calyx not only affords a more efficient protection to the flower-bud than a polysepalous one, but also gives support and protection to the base of the adult flower and to the developing fruit. Hence a gamosepalous calyx is never caducous.

In Umbelliferae, where the flowers are closely aggregated, and in many Compositae, where in addition they are surrounded by a ring of bracts, a protective calyx is to required and is either very small or quite absent. The calyx, however, may perform other functions. Thus, in many Compositae (e.g. *Taraxacum*, *Senecio*, and *Cnicus*) there is a rudimentary calyx represented by hairs, forming a silky pappus, which undergoes further development after flowering, and serves to disperse the fruit. In some flowers also the sepals instead of being green, as is usually the case, are brightly coloured, and attractive. In this case the calyx is described as petaloid. In a polysepalous calyx the outline of the individual sepals is described in the same terms as are used for ordinary foliage leaves. The number of sepals in a gamosepalous calyx is usually indicated by divisions or teeth. If the divisions pass almost to the base of the calyx, it may be described according to their number as 3-, 4-, 5-*partite*; if about half-way down, 3-, 4-, 5-*fid*; if the divisions are small, 3-,4-, 5-*toothed*.

COROLLA

It is in the corolla where the greatest diversity of shape, colour and arrangement occurs. In some flowers the petals are entirely absent, for example, the willow (*Salix* species). In many cases the petals join to form a tube, as in the primrose. The number of petals, too, varies considerably, even more than in the case of the sepals. For example, the wall flower has four, the pink (*Dianthus* species) and the buttercup, five, and some, for example, the white water-lily (*Nymphaea alba*), have an indefinite number. Some flowers are regular in the arrangement of their petals. For example, in the case of the butter cup or the wallflower, the petals and, indeed, all the floral organs are symmetrical about any axis. That is, it does not matter through what vertical plane the flower is cut, the two halves produced are the mirror images of each other. Such regular flowers are said to be *actinomorphic*. On the other hand, many flowers are irregular.

In the sweet pea (*Lathyrus odoratus*), for example, there are five petals but they are not all of the same shape. Looking straight towards the inside of the flower, there is one large petal standing up at the back. It is larger and more spreading than any of the others, and is called the standard. Then, there are two wing-like petals, one on each side of the standard. Each is called a wing. At the bottom are two still smaller petals, facing each other and appearing similar to a ship's keel. The two together are therefore called the keel. In this flower it is quite obvious that there is only one vertical plane through which the flower could be cut in order to produce two symmetrical halves. The plane would pass down the middle of the standard and between the two wings and the two petals forming the keel. Such irregular flowers as these are said to be *zygomorphic*. Violets (*Viola* species) and the white deadnettle (*Lamium album*) are also examples of zygomorphic flowers.

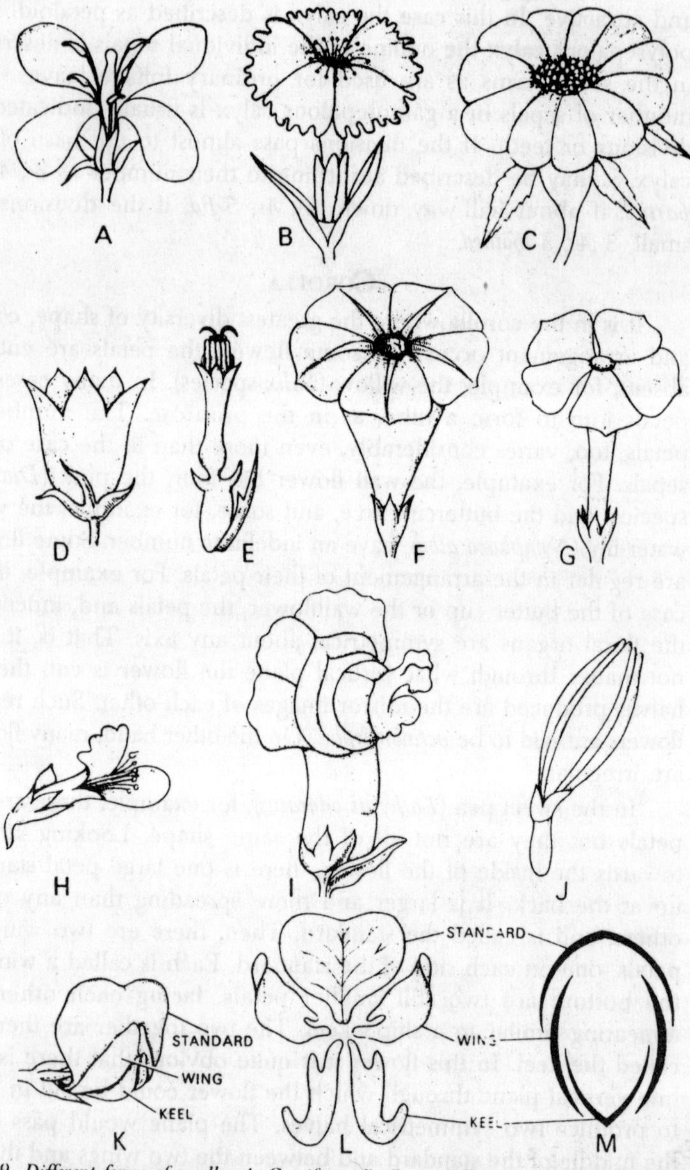

Fig. 1.9. Different forms of corolla. A–Cruciform, B–Caryophyllaceous, C–Rosaceous, D–Campanulate (Bell shaped), E–Tubular, F–Infundibuliform (Funnel shaped), G–Rotate, H–Bilabiate, I–Personate, J–Ligulate, K–Papilionaceous, L–Different parts of a papilionaceous corolla separated, M–Papilionaceous aestivation.

In a large number of flowers, it is impossible to distinguish a separate calyx and corolla. Good examples of this are the tulip, bluebell, crocus, lily, etc. There are, however, two whorls of brightly coloured segments, an outer one of three and inner one of three, which alternate with the three outer ones. In such examples, where there is no differentiation into sepals and petals, the two whorls are known collectively as the *perianth.* The corolla may consist of spirally arranged, free, non-coherent petals, e.g Cactaceae; or of a single whorl (*Digitalis, Malva*), or, more rarely, a double whorl of petals (*Papaver*). In water lily the petals are arranged in a close spiral.

The corolla may be *polypetalous* or *gamopetalous* (cf. calyx), *regular* or *irregular* and, as it to a large extent determines the symmetry of the flower, the terms *zygomorphic actinomorphic,* are applied to it. According to the insertion of petals the corolla is described as hypogynous, perigynous, or epigynous. The corolla serves, in most cases to attract insects to the flower in connection with process of pollination. It also protects the stamens and carpels, especially when the petals are united in the form of a tube enclosing these essential organs. The tube also serves as a receptacle for nectar.

After fertilization the seeds begin to develop, and the corolla is usually shed. The withered corolla, however, may persist in a few cases. The petals are usually brightly coloured, but sometimes green (sepaloid). They may be absent, e.g *Alchemilla* and some Ranunculaceae (*Clematis, Anemone*) or represented by nectar-secreting structures, e.g. *Aconitum* and *Helleborus.*

In a polypetalous corolla the outlines of the individual petals are described in the same terms as are used for the foliage leaf, and, as in the calyx, the gamopetalous corolla may be described as 3, 4, 5- partite, -fid, or toothed. The following special terms are applied to polypetalous corollas; *Cruciform* where the corolla consists of four clawed petals arranged crosswise, *i.e.* in the diagonal planes of the flower (e.g. mustard and Cruciferae); *rosaceous,* if it consists of five spreading petals, not clawed, and attached perigynously (Rosaceae); *caryophyllaceous,* if it consists of five clawed petals, with spreading limbs attached hypogynously to the receptacle inside a slender tubular calyx (pinks and many Caryophyllaceae); *papilionaceous* (from the supposed resemblance to a butterfly), if it consists of five petals, one large–the standard, two lateral–wings, and two fused to form a boat-shaped structure–the keel, *e.g.* pea and other members of the Papilionatae (Leguminosae).

The Corona

This is the term applied to the whole series of outgrowths developed on the corolla or perianth of certain flowers. In *Narcissus*, where the perianth is gamophyllous, the outgrowths are coherent, and the corona is cup-shaped. Ak or akanda *Calotropis gigantea* furnishes a good example of a corona.

Aestivation

This has already been referred to. The aestivation of the perianth (or calyx and corolla) only can be studied. The folding of the individual floral leaves is described in the same terms as are used for foliage leaves. The aestivation of calyx or corolla may be *valvate*, *imbricate*, or *contorted* (*twisted*). Aestivation may be recognized either by taking transverse sections of young flower-buds, or carefully removing the young floral leaves one after the other.

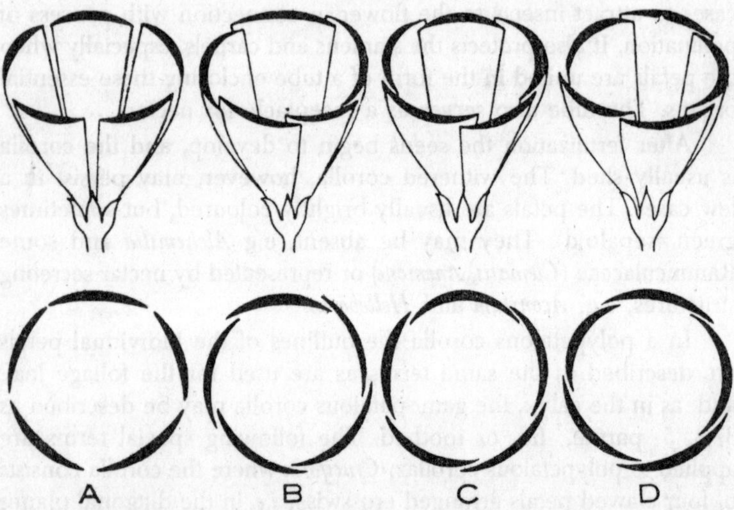

Fig. 1.10. Different types of aestivation of Calyx and Corolla. A–Valvate, B–Twisted, C–Imbricate, D–Quincuncial.

The Androecium

A stamen consists of three parts–filament, anther, and connective. The anther is usually two *lobed*, and contains microspores or pollen-grains. These lie in four cavities, the microsporangia or pollen-sacs of which there are two in each anther-lobe. The anther-lobes are connected by a strip of tissue containing a vascular bundle. This is the *connective*. It is usually narrow, and the anther-lobes lie

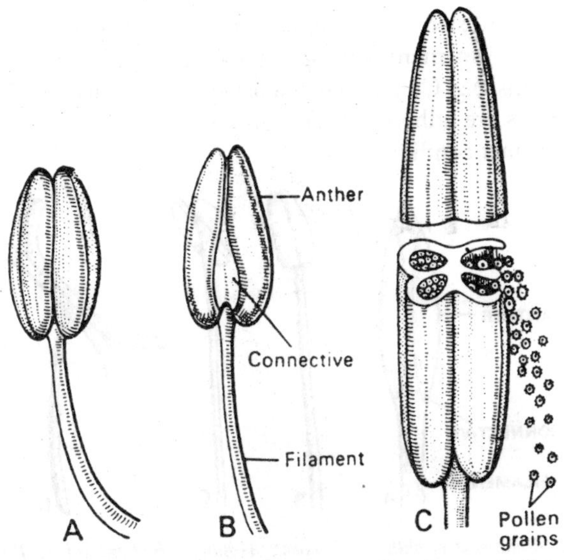

Fig. 1.11. Stamens showing parts. A–Ventral view, B–Dorsal view, C–Enlarged with T.S.

close together, but may be elongated so that the lobes are widely separated, as in some Labiatae. In a few cases (e.g. *Malva, Corylus, Carpinus*) the stamens when quite young undergo division or segmentation, and thus in the fully developed flower the anthers have only one anther-lobe with two pollen-sacs.

Sometimes special appendages are developed on stamens. These generally arise as outgrowths of the connective. In *Viola* there is a membranous orange-coloured outgrowth on top of anther, and , in addition, the two antero-lateral stamens have each a green elongated process (functioning as a nectar gland) passing down into the spur of the anterior petal. Barren or rudimentary stamens are called staminodes. They may consist only of filament or be represented by various modified forms. Some genera such as *Rosa* and *Prunus* which have "single" flowers in the wild state, possess "double" flowers in cultivated forms. If such "double" flowers are dissected it is frequently possible to trace a transition from stamens, and often from carpels also, to petals. Where doubling is complete the flower is sterile and such horticultural varieties are propagated vegetatively.

The stamens may be hypogynous, perigynous, or epigynous; but sometimes, owing to common basal growth, they adhere to the corolla (or perianth). They appear then to be developed on the petals, and are said to be epipetalous (*epiphyllous*, if on a perianth). This is found in many gamopetalous, or gamophyllous, orders of Angiosperms, e.g. Compositae, Labiatae, Convolvulaceae. Sometimes the stamens are adherent to the gynaeceum, e.g. in orchids; this is the *gynandrous* condition.

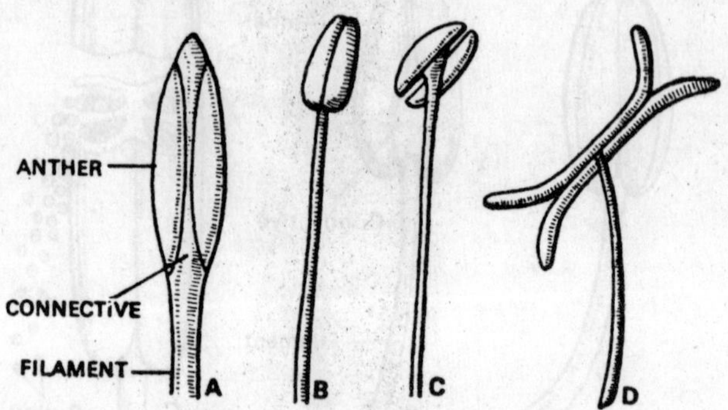

Fig. 1.12. Attachment of anther to filament. A–Adnate, B–Basifixed, C–Dorsifixed, D–Versatile.

If the stamens are free from each other, *i.e.* not coherent, the androecium is polyandrous (diandrous, triandrous, pentandrous, etc., according to the number). If united, the union may be of two kinds. (a) The stamens cohere by their filaments; this is the adelphous condition–monadelphous if all are united to form a tube

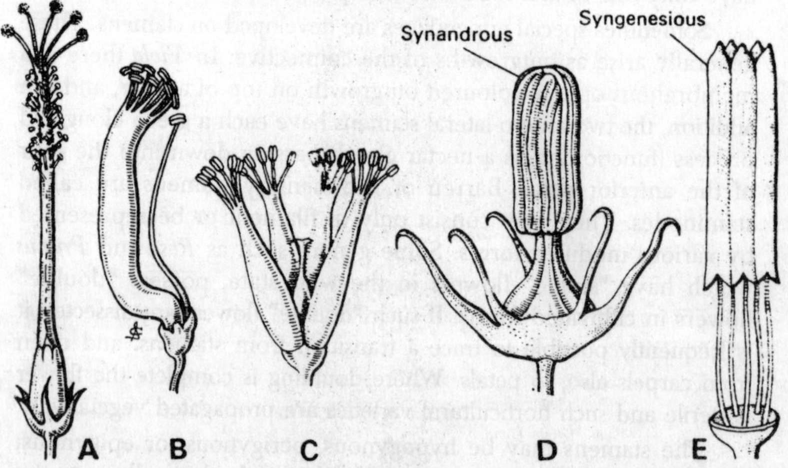

Fig. 1.13. Cohesion of Stamens. A–Monadelphous, B–Diadelphous, C–Polyadelphous, D–Synandrous, E–Syngenesious.

round the pistil, diadelphous if united in two groups, *Polyadelphous* if in several groups. The monadelphous condition is found, for example, in *Malva* and some Papilionatae; the diadelphous, in other Papilionatae (e.g. pea) where, of the ten stamens, nine are fused and the tenth is free; the polyadelphous, in Anonaceae and Rutaceae. (b) The stamens cohere by their anthers, the filaments being free. This is characteristic of Compositae, some Solanaceae (e.g. *Solanum* species), etc. It is the syngenesious condition. Where the stamens in a flower have different lengths, special terms are sometimes applied to the androecium. Thus, in the Family Cruciferae, there are frequently four long and two short stamens, and the androecium is said to be tetradynamous. In Labiatae and Scrophulariaceae where there are two long and two short stamens, it is didynamous. These are the only common orders in which these terms are used.

Insertion of the Anthers

The attachment of the anther to the filament should be noticed. It is *innate* or *basifixed* if the anther is fixed directly on top of the filament; *adnate* if the connective is well marked, and there is no articulation of the filament to the base of the anther, so that the filament seems to run up the back of the anther; *dorsifixed* if the filament is attached to the back of the anther and the anther is immovable; *versatile* if the attachment is similar, but the anther swings on the filament.

Development of Stamen

The stamen arises as a protuberance on the receptacle, and soon shows an external differentiation into anther and filament. At an early stage in the development of the anther, the two anther lobes appear. Further development can be followed by means of sections of progressively older material, which has been carefully fixed and stained, viewed under a microscope.

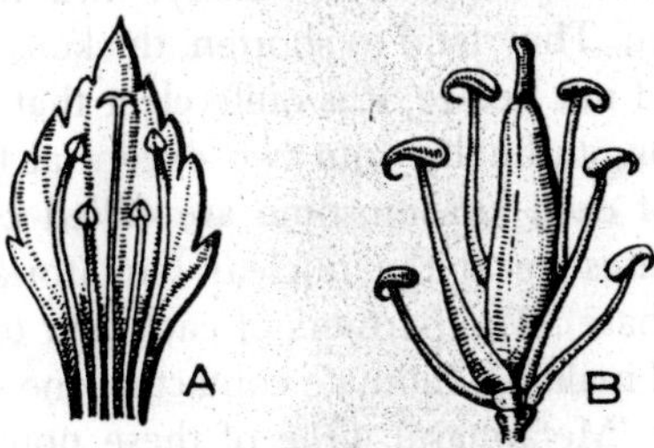

Fig. 1.14. Length of stamens. A–Didynamous, B–Tetradynamous.

Development of the Anther

In each anther-lobe two groups of hypodermal cells divide by periclinal walls, separating an outer layer of cells from an inner. The latter is the primary archesporium. The cells of the outer layer further divide by two successive periclinal walls to form three layers of cells. The outermost of these layers becomes the *fibrous layer* of the anther wall, and the inner most becomes the *tapetum*. The primary archesporium divides repeatedly to form ultimately the spore-mother cells. Each spore-mother cell, by tetrad division, gives rise to four *microspores* or pollen-grains. The process of nuclear division of the spore-mother cell to form four daughter cells is known as reduction division or meiosis. We have already seen that mitosis involves the resolving of the cell nucleus into a number of chromosomes.

Each chromosome splits longitudinally and the halves separate and move to opposite poles, where they associate together as two daughter nuclei. Thus in mitosis the daughter nucleus has the same number of chromosomes as the parent nucleus. In meiosis, on the other hand, each daughter nucleus has only half the number of chromosomes of the parent nucleus. The reason for this is that at the first division of the spore-mother-cell nucleus, whóle chromosomes move to opposite poles. This division is quickly followed by a second division corresponding to ordinary mitosis. These two division constitute the process of meiosis. The behaviour of the nucleus during meiosis is as follows.

The spore-mother-cell nucleus resolves itself into distinct slender threads, the chromosomes. These associate in pairs, but not at random., because it has been shown that one chromosome of each pair comes from the male, the other from the female parent of the preceding generation. The partners of the pair of chromosomes (*a* and *b*) is represented diagrammatically, and their behaviour subsequently traced. They tend to shorten, thicken, and inter-twine, and at this stage, if not before, it is quite clear that each partner of the pair is split longitudinally into two except at the centromere. The two halves of each chromosome are called chromatids. The paired chromosomes lie at the nuclear (equatorial) plate of the mother cell (Prophase). The partners of each pair tend to fall apart somewhat, but still maintain intimate contact in one or more regions along their length (Metaphase). One of these points of contact is the centromere.

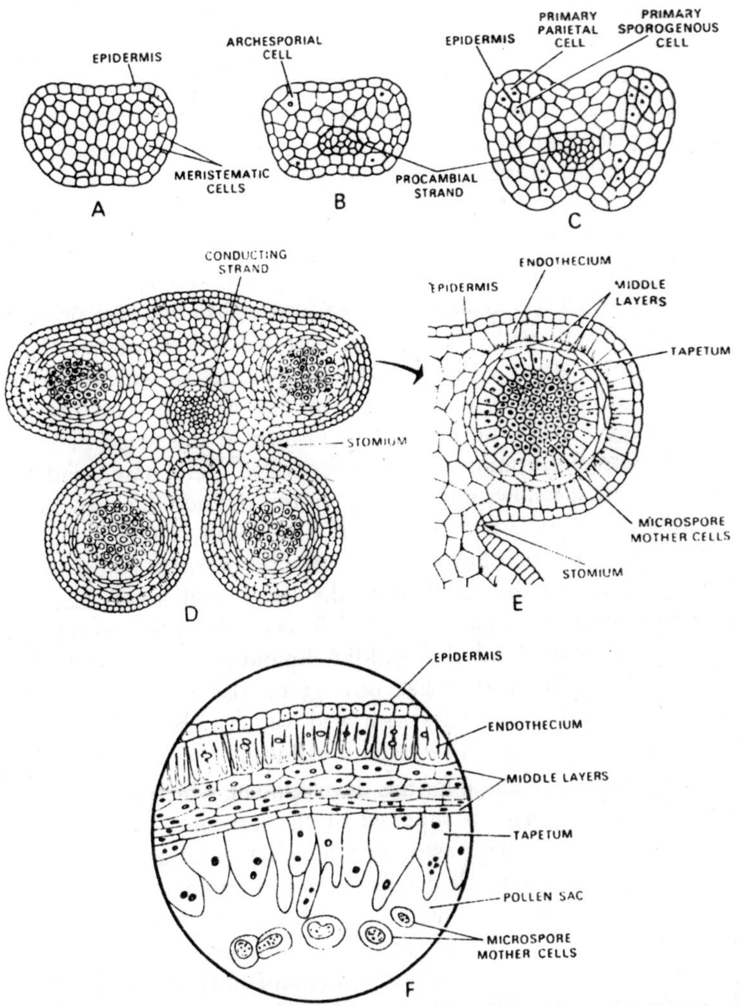

Fig. 1.15. Different stages of the development of anther. A–T.S. of young anther. B–Differentiated archesporial cells, C–Differentiated primary parietal cells and primary microsporangenous cells, D–T.S. of young anther showing four microsporangia and vascular bundle, E–Detailed structure of one microsporangium, F–A portion of microsporangium showing details.

Corresponding with other points of contract we may get a transverse division of two chromatids, one of each pair of chromosomes. These transverse divisions divide the chromatids into segments of corresponding length, and the points at which they occur are the chiasmata. From the centromere, forces operate which bring about the complete separation of the two chromosomes, except that in the process of being forced or pulled apart similar segments of two chromatids of the homologous pair of chromosomes may change places. Thus, the segment at the upper end of chromatid 3 changes place with the corresponding segment from chromatid. This process is called crossing over. It involves an interchange of nuclear material between two chromatids of an homologous pair of chromosomes.

The separated chromosomes (*a* and *b*) continue to move along a spindle which has meanwhile been forming, to opposite poles of the cell (Anaphase). If crossing-over has taken place, these chromosomes are not exactly identical with those which entered into association at the beginning. This has an important bearing on genetics. Usually no cell-wall is formed at this stage between the two sets of chromosomes thus separated (Telophase), and the second division follows immediately. In this division the chromatids separate, 1 from 2, and 3 from 4. The result is four daughter nuclei, each of which has only half the chromosome complement of the original mother nucleus of the spore-mother-cell. Each of these four nuclei is said to be haploid. This condition is conveniently represented by the symbol *n*, and the diploid condition by 2*n*. The spore-mother-cell nucleus has now given rise to four daughter nuclei. Each of these becomes associated with cytoplasm, and around each cytoplasmic mass is secreted a cell-wall.

The nutritive material for this development, which also involves expenditure of energy, is supplied by the tapetum. The four cells thus formed may remain associated together in tetrads, frequently forming regular geometrical figures such as triangular pyramids, but more usually they separate as individual microspores. Meanwhile the anther has been growing, and the microspores occupy the four cavities in most mature anthers. Simultaneously the anther wall develops and some differentiation takes place in it in relation to a dehiscence mechanism. This may take the form of a fibrous layer,

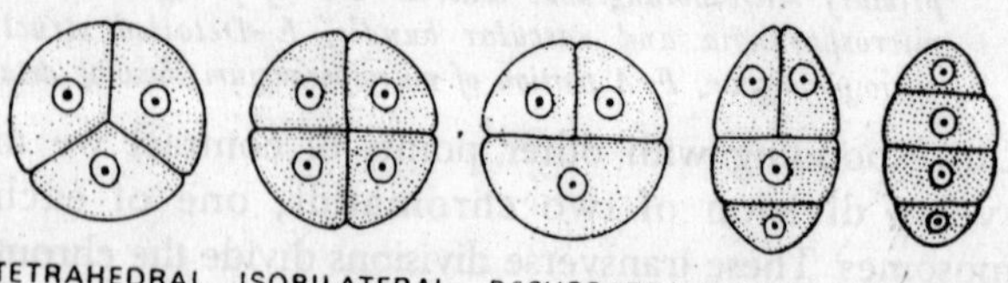

Fig. 1.16. Different types of microspore tetrads.

immediately below the epidermis, the cells of which have characteristically thickened walls.

Dehiscence of the Anther

In order that the microspores may be liberated to bring about pollination, the anther dehisces. When this takes place in many stamens a longitudinal split appears between each pair of pollen-sacs in each anther lobe. This split gradually enlarges by the separation and curling back of the anther wall on either side of it. The wall separating the two pollen-sacs also breaks down, and thus the pollen-sacs are exposed as one cavity. Two such cavities will therefore be formed in each anther, one on either side of the connective. If these cavities face towards the centre of the flower, the dehiscence is said to be *introrse*; if towards the outside, *extrorse*; if sideways, *lateral*. The splitting and curling of the anther walls is

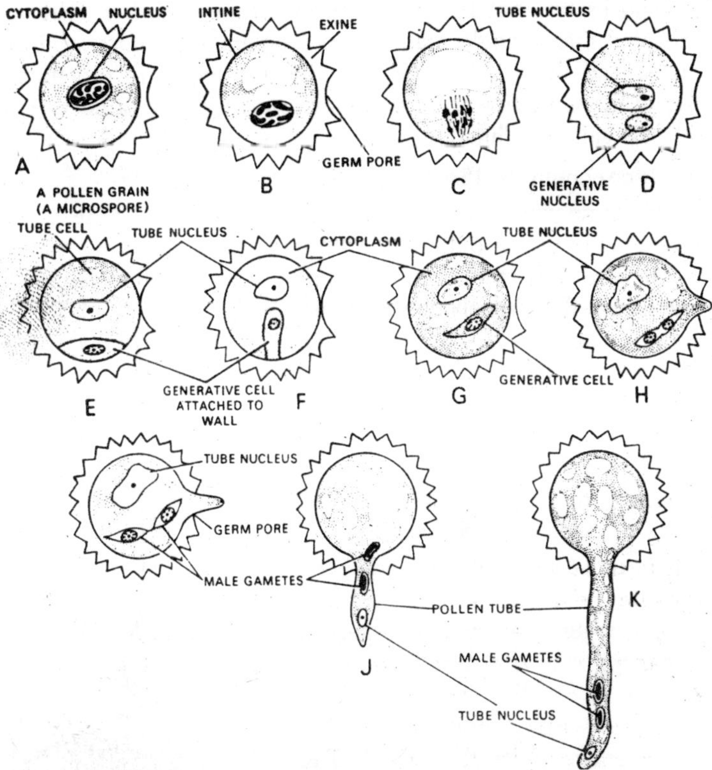

Fig. 1.17. Development of male gametophyte with the formation of male gametes.

brought about by the action of the fibrous layer. As these cells lose water by evaporation they contract, but the arrangement of the thickening of the cell-walls permits of contraction only in the plane parallel to the surface. The strain thus set up causes a break in the anther wall along a prepared line between the pair of pollen-sacs in each anther lobe.

The outer cell-walls of the cells of the fibrous layer contract at a much greater rate than the inner walls on account of the differential thickening. Hence the curling outwards and backwards of the anther walls. The microspores vary much in size, form and colour, in different plants. At first they are unicellular and the wall consists of two membranes. The outer coat, the *exine*, frequently has protuberances, spines, etc., and also one or more areas where the wall is thin and forms a pore. The inner coat, the *intine*, is thin and consists of cellulose. In some plants, e.g., *orchids*, the microspores are not loose, but those of each anther lobe are aggregated into a single mass called a pollinium. Before seed can be produced the microspores must be transferred to the stigma, either of the same flower, or of another flower of the same species; a process called pollination.

The Gynaeceum or Pistil

The gynoecium shows many interesting forms. One of the simplest is that of the buttercup, where all the carpels are separate. The number, too, is high and indefinite. In many flowers, on the other hand, the carpels are few and definite in number, and are very often joined to each other. The methods of joining of the carpels are interesting and varied. The best way to examine the various types would be first of all to imagine the carpel as an open leaf, bearing its ovules on the margins.

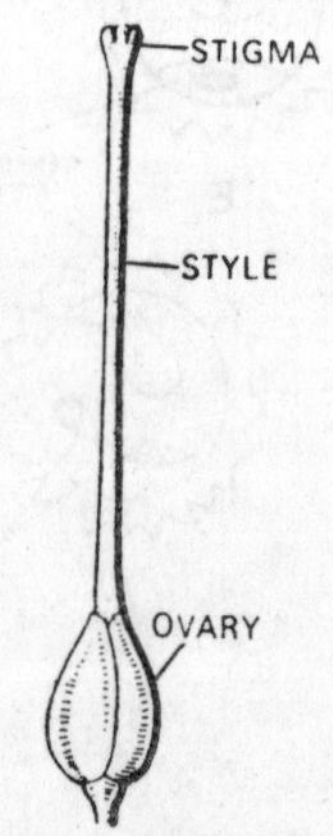

Fig. 1.18. Syncarpous gynoecium (Pistil) showing different parts.

It must be remembered, however, that ovules never are borne naked and exposed in this way in Angiosperms; but, starting with this *hypothetical* case, it is easy to see the various ways in which the ovules could conceivable become enclosed by the carpel. The simplest method for the carpel to enclose its ovules would be to fold in half, thus bringing together its two margins bearing the ovules. Then, imagine these margins to become fused. Thus one carpel enclosing a vertical line of ovules would be produced. In the bean, for example, there is a series of about eight ovules arranged as one would expect them in such a case. The next method of enclosure would be for two carpels, facing each other, to join at their margins. This would give two carpels forming an ovary with one common cavity into which the ovules would project in two longitudinal rows. This is the case in the gooseberry (*Ribes uva-crispa*). A variation of this is seen in the tomato. Here, the margins, four in all, two from each carpel, meet at a common centre. Thus there are two carpels, with two cavities, and a row of ovules projecting into each cavity.

In many other cases the same method as that of the gooseberry applies, except that there are three carpels, with a common cavity into which three longitudinal rows of ovules project. This is seen in the violet. Then a type similar to that of the tomato (*Solanum lycopersicum*) except that it has three carpels with three cavities, into each of which one row of ovules projects, is also possible; indeed, it is very common among flowers. It is seen, for example, in the tulip. A completely different type is seen in the primrose. Here, there are five carpels joined together; but the ovules, instead of forming rows at the fused carpellary margins, are borne on a projection from the base of the ovary cavity.

The gynaeceum or pistil consisting of megasporophylls or carpels forms the inner essential organ of the flower. It may consist of one or of several carpels. In the latter case, according to the number, it is bicarpellary, tricarpellary, etc. The old classical view represents it as a single carpellary leaf folded with the lower side outermost and its margins coherent along a line (the *ventral suture*), and, further, with the apex of the leaf elongated and slightly swollen at the tip. In such a monocarpellary pistil the hollow basal portion of the folded carpel is the ovary, from which the fruit is afterwards developed; it contains a varying number of ovoid or rounded bodies, the ovules or megasporangia which after fertilization develop into seeds. The slender prolongation, of varying length, on top of the ovary in the style, which usually contains a central cavity communicating with the cavity of the ovary, but may be composed of loose tissue throughout. The apical portion of the style, called the stigma, is usually swollen and covered with hairs or glandular

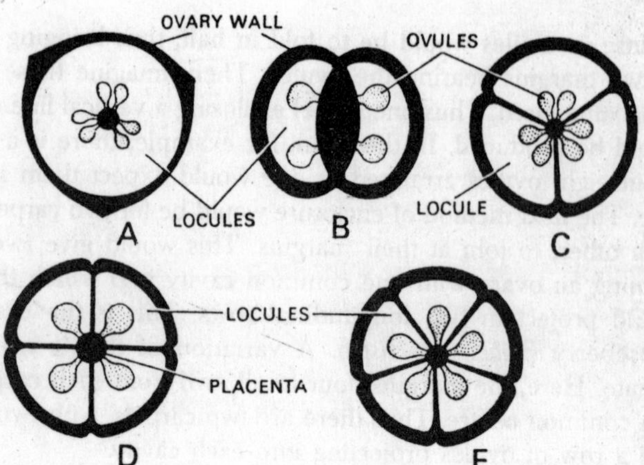

Fig. 1.19. T.S. of Gynoecium showing number of locules. A–Unilocular, B–Bilocular, C–Trilocular, D–Tetralocular, E–Pentalocular.

papillae; as we shall see later, it forms the receptive surface for the microspores. When there is no style, the stigma is said to be sessile on the ovary. If we examine the ovary, we find that the ovules are developed in two rows, each row on a longitudinal ridge or cushion of tissue called the placenta, along the ventral suture, on the inner surface of the ovary wall. The dorsal suture is opposite to it. The placentation (*i.e.*, the position or arrangement

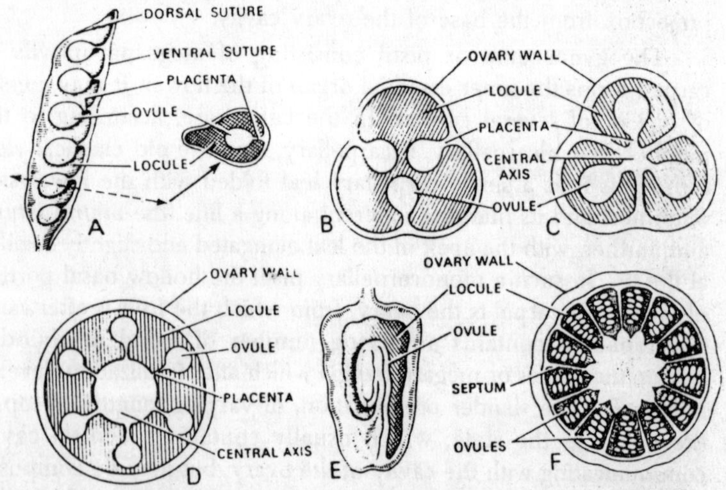

Fig. 1.20. Different types of placentations. A–Marginal (L.S.) and (T.S.), B–Parietal, C–Axile, D–Free central, E–Basal, F–Superficial.

of placentas in an ovary) in the *simple* ovary is described as marginal. The pistil of the Papilionatae (pea, bean, etc.,) is a good example of this. Although morphologists have compared the formation of the monocarpellary pistil to the folding of a carpellary leaf, we must not assume that this process can be observed during the development of the flower. But in certain Gymnosperms the ovules are not enclosed in an ovary, but are borne on the open megasporophylls. In *Cycas* they are borne on the margins of the megasporophylls. They imagined such a megasporophyll folded so as to bring the margins together and form a cavity enclosing the ovules, thus giving the condition found in such Angiosperms as the pea.

Polycarpellary Gynaeceum

Of this there are two conditions, according as the carpels are or are not united with each other. If the carpels are free, each forms a simple ovary, style and stigma, like the single carpel of the monocarpellary pistil. This is the apocarpous condition. Here, while there is a single gynaeceum or pistil in the flower, there is a number of simple ovaries. The number indicates the number of carpels. The placentation is marginal. Frequently, only one ovule is developed in each loculus (many Ranunculaceae and Rosaceae). In the second condition all the carpels are united to form a single *compound* ovary, and the pistil is syncarpous. The union may or may not be complete. If complete, the ovary bears a single style and stigma, and it is only by the internal structure of the ovary that the number of carpels can be determined. If incomplete, a number of styles or stigmas are borne on the single ovary, owing to the apices of the carpels remain in free.

With few exceptions the number of styles or stigmas is the same as the number of carpels. Hence, *e.g.,* in Compositae, where the style is single, but there are two stigmas, we assume that the pistil is bicarpellary. The structure of the ovary and the placentation in the syncarpous pistil differ in different cases. The following conditions should be noticed:

(a) If the carpels are fused by their adjacent margins they form a unilocular ovary. The fused margins are the placentas bearing ovules, and the placentation is marginal and parietal. The number of parietal placentas indicates the number of carpels.

(b) If the carpels are folded on themselves before fusing, that is, if the fused margins run in to the middle of the ovary, a multilocular ovary results and the marginal placentas of all the carpels fuse in the centre to form a central or axile column. The placentation is marginal and axile. The number of loculi, or the number of septa by which the ovary is divided, indicates the number of carpels (except where *false* septa are formed). Sometimes only one ovule is developed in each loculus.

In the ovary of poppy there is an intermediate condition between (a) and (b). The septa, which are covered with ovules and are therefore placentas, do not reach the middle of the ovary. The ovary is unilocular, but partially divided. The placentation is parietal.

(c) If the carpels fuse by their adjacent margins and the ovary is unilocular as in (*a*), and the ovules are not developed on the carpellary margins, but are borne on a central axis running through the middle of the ovary, the placentation is free-central. In typical cases (Myrsinaceae) the central axis is a prolongation of the receptacle into the ovary. The ovules are developed on the axis of the flower, not on the carpels. There are, however, a few families (*e.g.,* Caryophyllaceae) in which the free-central placenta appears to be derived from an originally axile placentation by the breaking down of the septa.

Basal placentation is a modification of typical free-central. Here a single ovule is inserted on the floor of the ovary. It is developed on the receptacle, which, however, is not prolonged as an axis into the ovary (e.g. Polygonaceae and Compositae). The ovules are, in rare cases, developed, not on the margins of the carpels, but over the whole inner surface, e.g., the flowering rush (apocarpous) and white water lily (syncarpous). This is called superficial placentation.

True and False Septa

True septa, or dissepiments, are those which represent the inturned margins of carpellary leaves. Septa formed in any other way, *e.g.,* as ingrowths from the surfaces of the carpels, are *false.* In the ovary of Cruciferae, for example, the false septum is formed by two membranes, which grow in from the two parietal placentas, and meet and fuse in the centre.

Superior and Inferior Ovaries

In all hypogynous and perigynous conditions, the ovary is described as *superior*; in the epigynous condition, as inferior. It might seem out of place to describe the ovary as superior, and the calyx inferior, in such perigynous conditions. But it must be realized that the ovary here is developed at the morphological apex of the receptacle and is free from the calyx.

Development of the Ovule

The development of the ovule is followed by examining series of longitudinal sections of ovules at progressively older stages of development. The material must be specially fixed and stained for this purpose. Such ovaries as those of pea, lily and marsh-marigold are useful material for examination, and if cut transversely at different stages of development the sections will show ovules cut more or less longitudinally. The young ovule arises as a small protuberance from the placenta and gradually increases in size and complexity. The body of the ovule is called the nucellus. Quite early in its development, one or two ridges of cell tissue appear at its base and in their further growth encroach over the still growing nucellus so as ultimately to enclose it, except for a small pore left at the apex which persists as the micropyle. The one or two coverings thus formed are the integuments. Meanwhile, from the basal region of the ovule, or chalaza, below the integuments, an elongation forms the funicle, or "stalk" of the ovule. This joins the chalaza at the hilum.

At an early stage in the development of the nucellus a hypodermal cell at the apex enlarges. It usually divides into two, an upper (outer) tapetal cell, and a lower (inner) sporemother cell. The nucleus of the latter divides twice, the division being meiotic, and cell-walls are formed. Of the four haploid cells thus resulting, only one is a functional megaspore. The remaining three show no further development. Exceptionally, as in *Casuarina*, all four are functional. The functional megaspore enlarges as the ovule grows, and comes to occupy the greater part of the nucellus.

Meanwhile, its nucleus divides into two daughter nuclei, one of which migrates to the micropylar, the other to the chalazal end of the megaspore. In these positions each nucleus divides further, giving rise to four nuclei at each pole. Three of the nuclei at the micropylar pole become surrounded by cytoplasm and form the two synergidae and the oosphere or egg-cell (female gamete). Three of the nuclei at the chalazal pole become surrounded by cytoplasm and cell-walls, and form the antipodal cells. The remaining two polar nuclei, one at each pole, migrate towards the centre of the megaspore, where they fuse to form the secondary nucleus, often

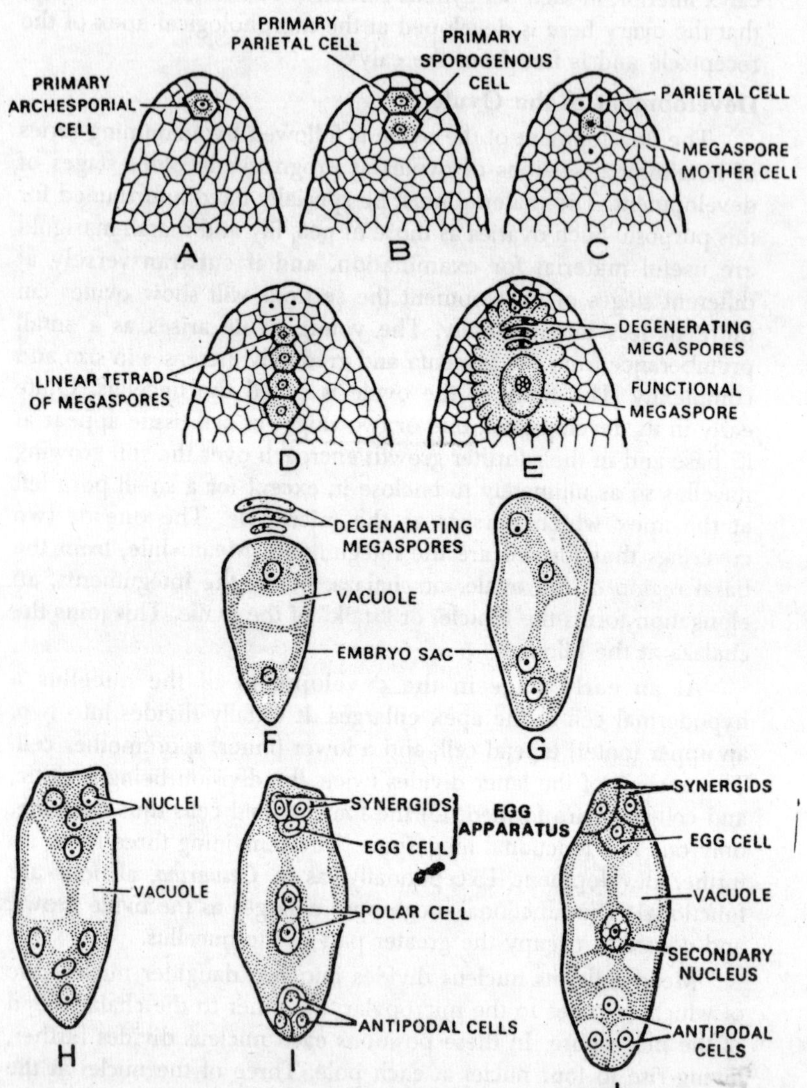

Fig. 1.21. A-E–Different stages of megasporogenesis, F-J–Development of female gametophyte (Embryo-sac) megagametogenesis.

referred to as the primary endosperm nucleus. Before cytological work revealed the true nature of the megaspore, it was known by the name embryo-sac, and it is still frequently referred to by this name.

Structure of the Mature Ovule

The structure of the ovule of *Caltha* at the time when it is ready for fertilization. The slender funicle attaches the ovule to the basal placenta of the ovary and carries a vascular bundle to the chalaza. The nucellus is invested by the two integuments, and the micropyle is seen as a narrow passage or pore at the apex. Almost filling the nucellus is the enlarged megaspore (or embryo-sac). Within the megaspore, at the micropyler end, are the three protoplasts of the egg-apparatus; one is the oosphere, the other two the synergidae. At the chalazal end are the three antipodal cells enclosed in cell-walls. In the centre of the megaspore is the secondary nucleus.

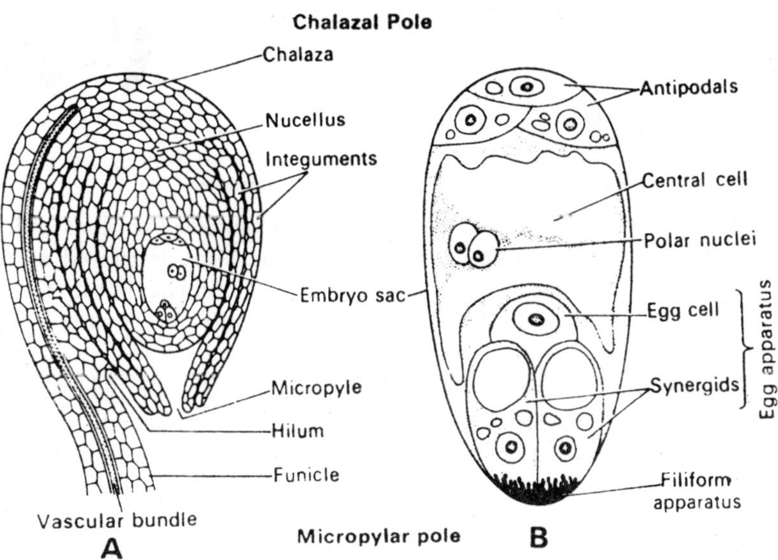

Fig. 1.22. A–Detailed structure of the L.S. of an anatropous ovule, B–Detailed structure of one embryo sac (female gametophyte)

Forms of Ovule

There are several forms of ovule to be noticed. One is the orthotropous ovule. Here the ovule is straight, not curved or bent appreciably. The chalaza and hilum lie close together, and the micropyle is at the extreme apex. In the *inverted* or anatropous form the body of the ovule has bent over during development, and fused for some distance with the funicle. This fused portion of the funicle is called the raphe. In this form the micropyle and

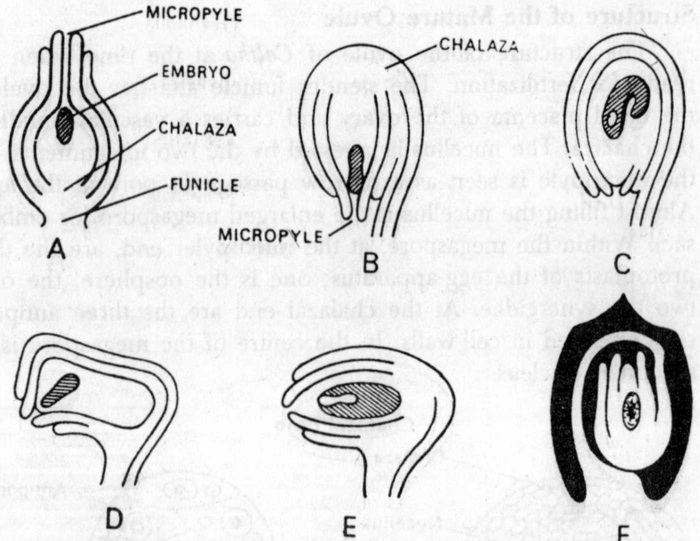

Fig. 1.23. Different types of ovules. A–Orthotropous, B–Anatropous, C–Amphitropous, D–Campylotropous, E–Hemitropous, F–Circinotropous.

hilum lie close together, and the chalaza is towards the other end. In the curved or campylotropous form the body is curved and bent round, so that the micropyle lies near the funicle; but there is no fusion with the funicle. Hilum, chalaza, and micropyle all lie close together.

The amphitropous ovule is an intermediate form in which the body of the ovule is straight, but has been twisted round, so that its long axis is at right angles to the funicle. Of these forms the anatropous is most frequently met with. Examples of the campylotropous ovule are found in many Cruciferae and Papilionatae (pea, bean, etc.). The orthotropous ovule is less frequently found, e.g. *Polygonum*. The Loganiaceae and some Cruciferae give examples of the amphitropous ovule.

Cohesion and Anhesion

The student must be clear as to the meaning of these terms. *Cohesion* is union between members of the same series of floral leaves. Thus gamosepalous, polysepalous, polyandrous, syngenesious, apocarpous, syncarpous are terms signifying cohesion or want of cohesion. *Adhesion* means union between members of different series, as when the stamens are epipetalous. We have already explained that cohesion or adhesion of parts in the flower is not due to the actual fusion of parts originally separate, but to common basal growth during development.

Vertical Sections and Floral Diagrams

The general structure and arrangement of parts in a flower may be shown in drawings of longitudinal or vertical sections (half-flower), in floral diagrams and floral formulae. The floral diagram may be described as a ground-plan of the flower showing the relation of the parts to each other and to the mother-axis. In making a floral diagram cohesion of parts may be indicated by connecting lines, and this may also be done in the floral formula which should accompany the diagram. An *empirical* diagram is one showing only the relative positions of the parts actually present. A *theoretical* one indicates, as well, by conventional means, the relative positions of parts which we assume were originally present, by are now lost.

The floral formula, together with the diagram and longitudinal section, enables us to represent the essential *morphological* features of the flower without a word of description being necessary. The symbols ⊕ and † respectively denote radially and bilaterally symmetrical (zygomorphic) flowers, the direction of the arrow indicating the plane of symmetry along which the flower can be divided into equal halves. The signs ♂, ♀, ☿ respectively denote staminate, carpellary, and hermaphrodite ("perfect") flowers. The letters K, C, and P represent calyx, corolla, and perianth, A and G the androecium (stamens) and gynaeceum (pistil), and the figure following each letter gives the number of parts in each series. Cohesion is indicated by brackets enclosing the number of parts; a horizontal bracket indicates adhesion between the parts of successive whorls; a horizontal line above the number after G means that the ovary is inferior, a line below, that it is a superior; the symbol is used when there are numerous parts in any series.

Thus the typical floral formula of Myrsinaceae

$$☿\ \oplus\ K(5)\ \overparen{C(5)\ A5}\ G(5)$$

reads–hermaphrodite, radially symmetrical flower, gamosepalous calyx of five sepals, gamopetalous corolla of five petals, androecium of five free epipetalous stamens, syncarpous pistil of five carpels with a superior ovary.

Modifications of the Flower

Although all flowers follow the same structural plan, they exhibit numerous modifications in design. Certain of these modifications increase the chances of pollination. Furthermore, they often reveal evolutionary relationships among plants, thus permitting botanists to place related forms together and to classify flowering plants according to descent, from the most simple to the most advanced and specialized forms.

Number and Arrangement of Floral Parts

The parts of the flower are usually arranged in whorls, or circles. These whorls are either four or five in number : one of sepals, one of petals one or two of stamens and one of carpels, united into a compound pistil. The number of parts in each whorl varies with the species, but it is usually constant and small. The two large subclasses of angiosperms (flowering plants) are distinguished by the number of floral organs in each whorl. In the dicotyledons, the number is four or five or multiples of these numbers.

A flower may have, for example, five sepals, five petals, ten stamens, and five carpels. However, the number of carpels in a compound pistil may be only two. In the monocotyledons, the number of organs in a whorl is three or multiples of three. In the

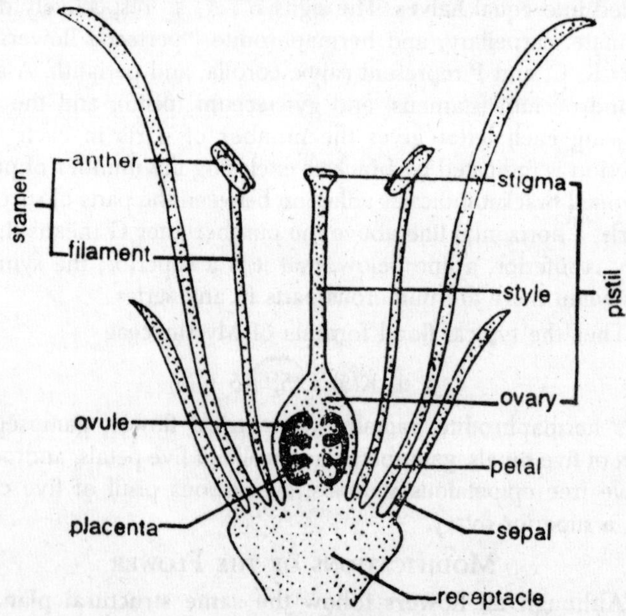

Fig. 1.24. Diagram showing the floral organs.

tulip, for example, there are six perianth parts, six stamens, and three carpels. In both subclasses, however, certain families have numerous stamens and carpels–generally more than ten. Among the families in which this condition occurs are the magnolia, buttercup rose, and waterlily of the dicotyledons, and the water plantain family of the monocotyledons. In some of these families the numerous stamens and carpels are attached to the receptacle spirally rather than in whorls. This combination of spiral arrangement and numerous stamens and carpels is considered indicative of a more primitive stage in evolutionary development than the whorl arrangement with the parts reduced in number.

Fusion of Floral Organs

Among the important modifications found in the flower are fusions of the organs of the same whorl and of different whorls. The dicotyledons, for example, are divided into two large groups, based upon the fusion of the petals. In one group the petals are separate. In the other group, the petals are joined to form a *corolla tube*, which is lobed at its margin. The lobes usually correspond to the number of petals. The sepals are usually distinct, but the basal edges are often joined into a shallow tube. The stamens are usually

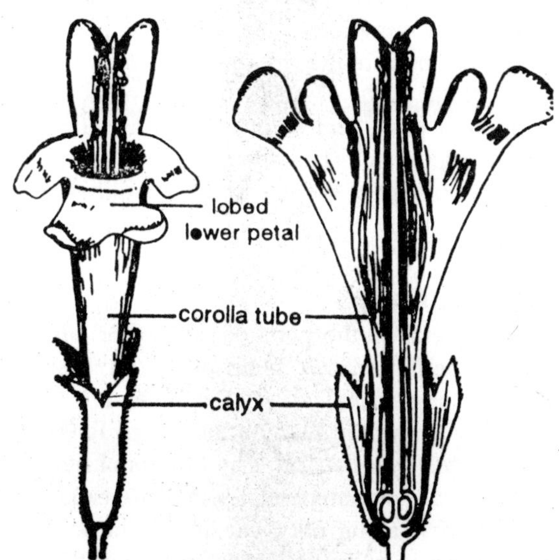

Fig. 1.25. Flower of a mint, ground ivy (Glecoma bederacea), illustrating fusion of floral parts and irregular symmetry of the flower.

free from one another but may be joined by their filaments or by their anthers into one or several groups. Sometimes free stamens are fused to the corolla tube at the base. The carpels, as we have seen, are commonly united into a compound pistil. In a flower with free parts, the sepals, petals, and stamens are attached to the receptacle beneath the ovary. Since the ovary is situated above the zone of attachment, it is said to be *superior.* This is the condition in a very large number of flowering plants.

In plum, cherry, blackberry, and other species, the basal portions of the sepals, petals, and stamens are united to form a cup-shaped *floral tube* (also called calyx tube). The lobes of the sepals and petals extend from the margins of this tube together with the unfused portions of the stamens. The ovary is still superior in this type of flower. One other type of fusion is that of the floral tube with the ovary. Then the sepals, petals, and stamens apparently grow from the top of the ovary, and the ovary is said to be *inferior.*

Actually the pistil is still the uppermost structure of the flower, just as in flowers with superior ovaries. The ovary appears to be inferior because of its union with the floral tube. Inferior ovaries are found in such plant families as the carrot, bluebell, sunflower, evening primrose, iris, cactus, and orchid. The position of the ovary, whether inferior or superior, is significant in plant classification and also helps in interpreting the nature of many kinds of fruits. The fusion of floral parts–of petals to form a tube, of carpels to form a compound pistil, and of ovary wall and floral tube–is an evolutionary development. The various organs do not become fused during the development of the flower but are united from the very beginning of growth in the embryonic cells at the apex of the floral axis.

Symmetry of the Flower

In many flowers, all the parts of each whorl are alike in size and shape, and the flower may be divided into two similar parts in more than one longitudinal plane. Examples are tomato, apple, petunia, morning-glory, and many others. Such flowers are said to be *regular* or *radially* symmetrical. The flowers of other species are *irregular*, or bilaterally symmetrical, for they are capable of division into two equal parts along only one longitudinal plane. In such flowers, the corolla is commonly more modified than other floral parts. In the garden pea the one top petal, the *standard*, is much enlarged. Two lateral ones, the *wings*, are much smaller, and the

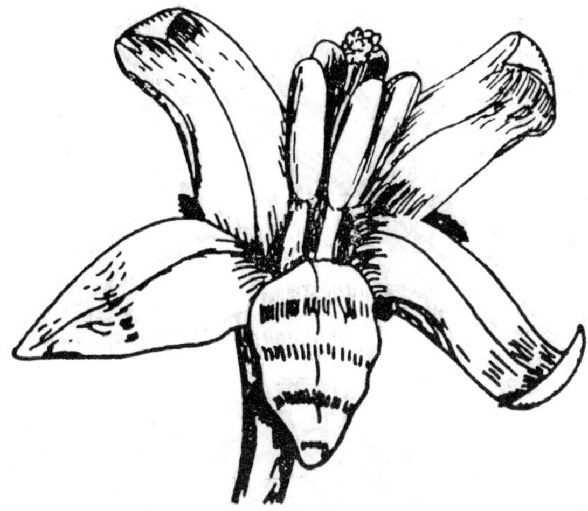

Fig. 1.26. Radially symmetrical flower of common nightshade (Solanum nigrum).

two remaining lower petals are united along one edge to form the *keel*, which encloses the stamens and pistil. Nine of the ten stamens are united a tube which surrounds the pistil. The tenth stamen is free. Other examples of irregular flowers are the snapdragons, mints, violets, and orchids. Irregular flowers are usually pollinated by insects, commonly bees.

Reduction in the Flower

The flower in which four readily distinguishable kinds of floral organs occur–sepals, petals, stamens, and pistils–is not the only kind. The flower may become reduced so that some organs are no longer evident, although they may be present in a vestigial form which can be recognized only after careful investigation. Plants in which the perianth parts have become minute or inconspicuous, or have disappeared, include the grasses and many trees such as willows and poplars, maples, oaks, hickories, and elms. Contrary to popular belief, a conspicuous perianth is not an essential feature of the flower. Reduction in the number of parts may also extend to the stamens and pistils. A flower in which both of these organs are present is said to be *perfect*, or *bisexual*. If one of these organs is missing, the flower is *imperfect* or *unisexual*.

Unisexual flowers are termed *staminate* if they lack a pistil or *pistillate* if they lack stamens. Both kinds of unisexual flowers may be present on the same plant, as in corn, most begonias, squash, pumpkin and cucumber, cattail, calla lilies, and many woody plants,

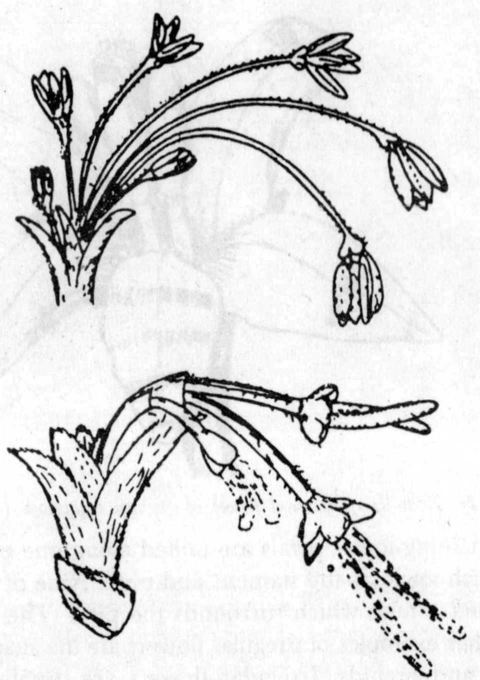

Fig. 1.27. Reduced unisexual flowers of box elder (Acer Negundo). Top–Staiminate flowers. Bottom–Pistillate flowers. The perianth is greatly reduced. This species is dioecious, for the two kinds of flowers are borne on different trees.

such as oaks, beeches, birches, walnuts, and hickories. Such plants are called *monoecious.* In corn, the staminate flowers are borne on the branched tassel at the top of the plant. The pistillate flowers are attached to the axis of the ear, or "cob", and are surrounded by the husk.

A single pistillate flower consists of an ovary to which is attached a thread-like silk 4 to 12 inches long. The silk bears numerous hairs which catch the pollen grains. The hairs are found even within the portion of the silk covered by the husk but the most numerous near the tip. Each thread of silk thus serves as a combined style and stigma. In *dioecious* plants the staminate and pistillate flowers occur on different plants of the some species. Examples of this condition are found in ashes and maples, in spinack, tree of heaven (*Ailanthus*), asparagus, Indian hemp, the date palm, hops, willows, holly, and poplars. The date palm probably represents the earliest recognized example of the dioecious condition,

Fig. 1.28. The monoecious condition in squash. Left–Staminate flower. Right–Pistillate flower. Both types are borne on the same plant.

for records show that the tree was cultivated in Mesopotamia as early as 3500 B.C. The farmers of this region recognized that there were male and female trees and practiced artificial pollination by bringing clusters of staminate flowers in contact with the pistillate flowers. The methods of bringing about fruit production in the date palm today differ but little from those of ancient times.

Staminate trees to provide pollen are grown with the pistillate trees, which bear the fruit. Propagation is achieved by planting offshoots from the base of the parent tree. The sex of the tree and the quality of the fruit which the pistillate plant will bear are thus determined in advance, for the offshoots always reproduce the parent type. Pollination is still done by hand, and several methods are used. A common procedure is to invert several of the long staminate clusters among the clusters of pistillate flowers. The pollen is shed and drifts to the stigmas of the pistillate flowers.

Floral Evolution

A number of theories have been advanced to explain the evolutionary origin of the floral organs. According to the most widely held of these, the flower is a highly modified axis, bearing appendages–the perianth parts, stamens, and carpels. It is comparable to the leaf bearing shoot. The internodes are suppressed above the level of the lowest sepals, and so the nodes are brought very close together. The floral organs, the sepals, petals, stamens, and carpels, exhibit many similarities to leaves in external form or internal structure and are generally regarded as homologous with leaves–that is, having had a common origin. The sterile appendages of the floral axis, constituting the perianth, show a greater similarity to leaves than do the uppermost, fertile appendages, the stamens and carpels. Nevertheless, none of the floral parts is believed to be descended directly from photosynthetic organs; they are thought to have merely evolved along similar lines, undergoing many transformations during their evolution. The carpel is thus interpreted as a modified leaflike organ folded inward along the midrib. Its margins are considered to have grown together, enclosing the ovules.

The apical region of the appendage has become modified into the stigma, which receives the pollen. This interpretation of the nature of the flower has led to the application of the term *sporophyll* (literally, spore leaf) to the stamen and carpel, since certain structures essential to seed formation, known as *spores*, are located within the ovule and anther. The flower is therefore defined as a group of a sporophylls, usually accompanied by a perianth. The view that the parts of the flower are transformed leaf like organs has met with considerable criticism in recent years, and the problem of the essential nature of the flower, from an evolutionary viewpoint, must be regarded as yet unsolved. Nevertheless, there is sound evidence that the flowers of certain plants, such as the buttercups and magnobas, are relatively primitive in structure and represent types from which other forms may have arisen.

Anatomy of Simple Flowers

Examples of anatomically (and primitively) simple flowers provide a basis for the understanding of more complex types.

Aquilegia

The stele continues in vestigial form beyond the level of departure of the uppermost appendages. The pedicel has five bundles which unite at the base of the flower, forming an unbroken cylinder. The sepal traces are then given off, three to each sepal, the three together leaving one gap. The traces to the petals, one to each, are next given off just below the level at which the sepal gaps close. Then the stamen traces pass off, one to each organ, in several whorls. Above the uppermost whorl of stamens the vascular cylinder becomes complete again. Then the dorsal traces of the carpels are given off, followed by the pairs of ventral traces to each carpel. These ventral traces are derived from the sides of the gaps formed by the dorsal traces, and their gaps therefore merge with those of the dorsal traces. The ventral traces immediately twist, passing out

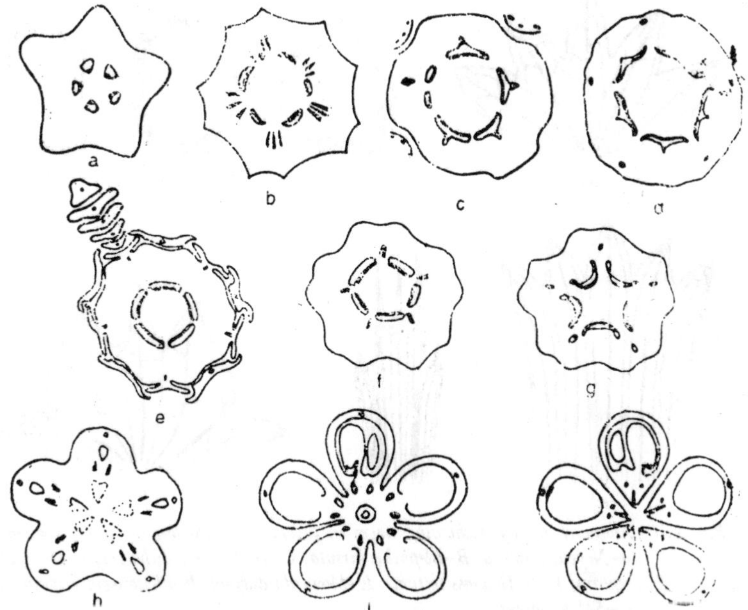

Fig. 1.29. Diagrams representing vascular structure of flower of Aquilegia as seen in series of transverse sections from pedicel to level above top of receptacle.

inverted into the carpel. Figure shows the condition in the receptacle just above the outward passage of the ventral carpel traces. The base of the loculus of each carpel is seen with the dorsal trace at the outside and the pair of ventrals on the inside. Internal to the ventrals the remainder of the vascular cylinder is seen as five masses of less strongly developed tissue (shown by dotted lines), chiefly phloem, with some weak xylem. This tissue becomes smaller and weaker toward the rounded top of the receptacle and fades out. (This fading out of vascular tissue is similar to that often seen in the vascular cylinder of determinate leafy stems.) Figure shows the carpels nearly free from the receptacle, with their three-bundle supply. Figure (above the receptacle) shows cross sections of the typical follicles with their dorsal bundles, and their pairs of inverted ventral bundles.

Anatomy of More Complex Flowers

Fusion in the Floral Skeleton

In the specialization of the flower, fusion and reduction of appendages have been the outstanding changes. Stages of *cohesion–*

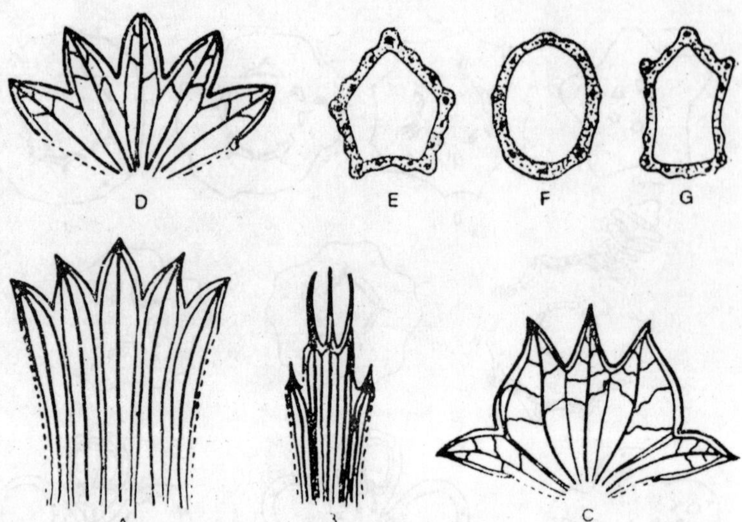

Fig. 1.30. Anatomy of calyces showing stages in fusion of lateral bundles under cohesion, A–Nepeta veronica; B–Blephilia hirsuta; C–Trichostema dichotomum; D–Ajuga reptans; E, F, G (cross section): E–Monarda didyma; F–Physostegia virginiana; G–Salva patens.

where members of the same whorl are fused to one another–and of adnation–where members of a whorl are fused to members of a whorl above or below–are found in many families. Evidence of reduction is to be seen in vestigial structures and through comparative study of related forms. In the evolutionary history of fusion, organs are at first united externally, and internal structure in not modified. The line of fusion is histologically obscure or absent; the vascular tissues are unaffected. Ultimately the fusion becomes more intimate and, wherever the vascular bundles lie close to one another, fusion may involve them also. Fusion may be between the traces alone or between the traces and the bundles within the appendages, but it seldom extends throughout the length of the bundles. Often there is ontogenetic and histological evidence of the fusion.

Fusion under Cohesion

The effect of cohesion on the vascular skeleton is chiefly upon lateral traces and bundles that are brought close together by gamosepaly, gamopetaly, syncarpy, and the marginal fusion of the upfolded margins of the carpel. It can be simply show by examples. The gamosepalous calyces of the mints shown all stages of fusion

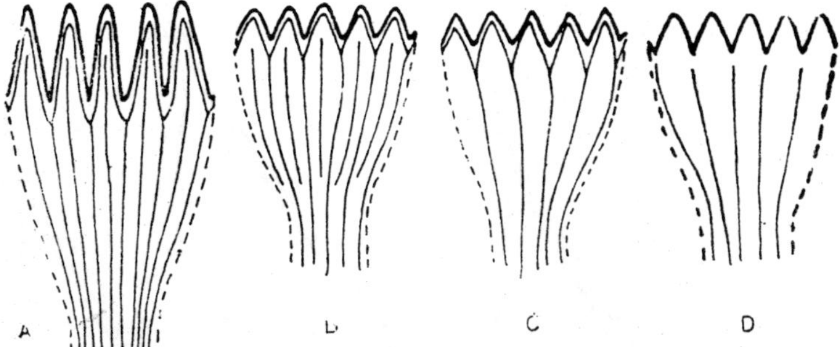

Fig. 1.31. Vascular system of gamopetalous corollas showing fusion of lateral petal-bundles and loss of bundles in reduction. A–Helianthus divaricatus; B–Senecio Fremontii; C–Chrysanthemum Leucanthemum; D–Xanthium orientale.

of the vascular tissue. In *Nepeta* each sepal has a midrib and two lateral bundles; there is no fusion. The calyx of *Ajuga* shows the pairs of adjacent laterals fused nearly to the sinuses. The zygomorphic calyx of *Blephilia* shows two pairs of laterals fused and three pairs free; that of *Trichostema* a similar condition, with the free pairs fused slightly at the sinuses.

A taxonomic character, number of calyx ribs, sometimes used in generic distinction in the Labiatae, is based on the extent of this fusion of lateral bundles. Examples are seen in *Monarda*, with 15 bundles, no fusion; *Physostegia*, with 10 bundles, all pairs of laterals fused; and *Salvia*, with 13 bundles, two pairs of laterals fused. In petals and carpels similar fusion is common. The folding of the carpel presents the simplest type of fusion under cohesion in the gynoecium. The two ventrals are brought close together, forming the pair of bundles characteristic of the ventral side of the primitive carpel, the follicle.

In earliest phylogenetic stages the carpel margins merely touch or are lightly fused. (The carpel is even not completely closed in some genera). With closer fusion, the epidermal layers on the surface of contact are lost and the two bundles form a double bundle; with complete fusion they become a single bundle that shows no histological or ontogenetic evidence of its doubleness. Fusion in the vascular tissue of a carpel may be present in the ventral bundles from an origin as one trace throughout their length, or may exist only in part of the carpel; where the ventral bundles arise as separate traces, they may unite at any point in their course. A carpel with two bundles, a dorsal and a ventral that consists of two

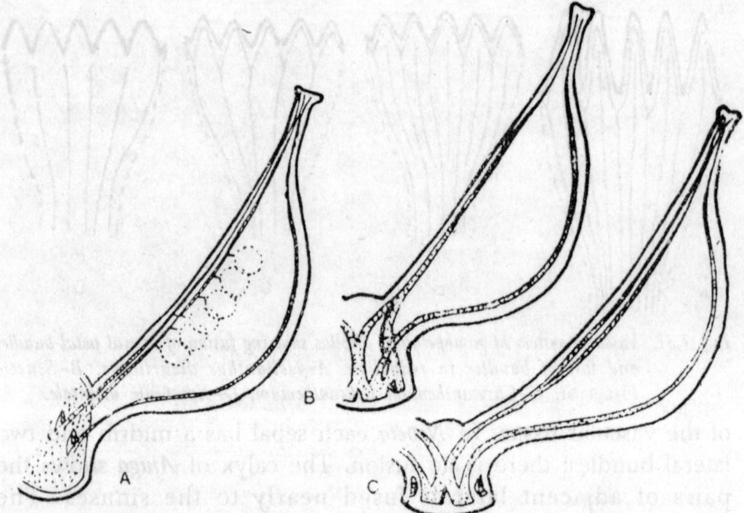

Fig. 1.32. Vascular supply of the carpel. A–three traces from three gaps, the bundles within the carpels unfused to the stigma; B–three traces from a dissected stele, the ventrals soon fused, only two bundles in the carpel; C–the ventral traces arising fused but becoming free in base of ovary. In B and C, the top of the floral stele is used up in the formation of the ventral traces.

bundles, may have one or two traces. Further, in much reduced carpels, especially in achenes, all three carpel traces may arise as one and separate only at the base of the loculus.

In any carpel, fusion may also occur among the bundles in the distal part of the ovary or in the style, when the traces and bundles are elsewhere free. Syncarpy involves fusion changes similar to those in free carpels. The lines separating the carpels and their margins are lost. The ventral bundles, inverted, form a ring of bundles in the center. They usually lie in pairs, each pair consisting of the ventrals of the same carpel, or, more often, of bundles from each of two adjacent carpels. The members of a pair tend to fuse and all degrees of fusion are found. In the center of a syncarpous ovary with three carpels, there may be a ring of six or of three ventral bundles–if of three, each bundle is morphologically double and represents either the two ventral traces of one carpel or one from one carpel plus one from the adjacent carpel. Where syncarpy arose while the carpels were still open, fusing margin to margin and enclosing a common loculus–or where the septa have been reduced or lost, and placentation is parietal–the ventrals of adjacent carpels lie in pairs, or fused, in the outer wall of the ovary. Carpel limits in such ovaries run through the placentas, which are morphologically double structures, each half belonging to a different carpel.

Fusion under Adnation

Whenever members of one whorl become fused to those of other whorls–whether the whorls consist of like or of different organs–their vascular bundles tend to fuse, as under cohesion. Fusion takes place between bundles that are near together either radially or tangentially and may involve any number of bundles belonging to two or more whorls. These stages can be seen in many families, the Crassulaceae, for example. Fusion involving several whorls, with union extending varying distances from the trace origin, is shown for Rosaceous flowers.

The Inferior Ovary

It is apparent that an ˙ :rior ovary can be formed by the adnation of the sepals, and stamens to the carpels or by the sinking of the gynoecium in a hollowed receptacle, with fusion of the receptacle walls about the carples. From anatomical evidence, it is clear that the inferior ovary in nearly all families is made up of appendages fused to the carples. In the ovary walls of some plants, *Alstroemeria*, *Hedera*, for example, the traces of all adnate organs

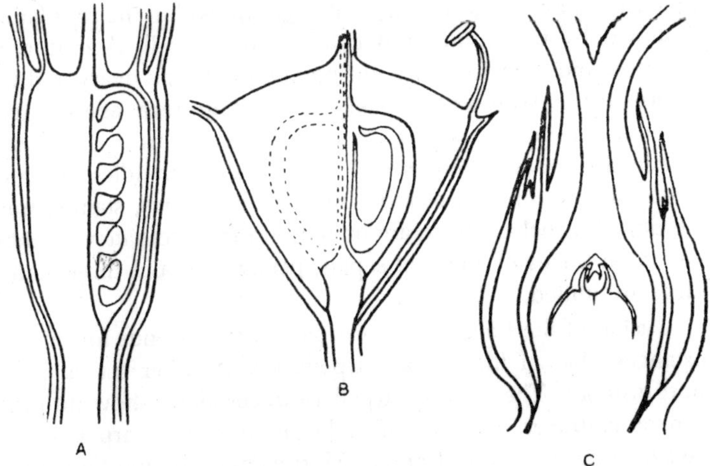

Fig. 1.33. Diagrams of vascular structure in inferior ovaries. A–Alstroemeria, B–Hedera helix and C–Juglans nigra.

run free to the base of the ovary in the positions they would have if there were no adnation. There is no fusion. In most species, however, with inferior ovaries, there is fusion within the ovary wall for various distances, as in *Juglans* and some of the Ericaceae.

The "ovary wall" in such forms, as is shown by the position and course of the bundles, clearly consists of the true ovary wall plus the tissues of the adnate organs. The sinking of the gynoecium in the receptacle necessarily involves invagination of the top of the stele. An "ovary wall" so formed contains two cylinders of stelar bundles, the inner inverted. Within the cavity, the distal part of the receptacular stele, from which the carpel traces are derived, turns downward and the morphologically uppermost carpels are the lowest. *Darbya* shows this well for the Santalaceae, the only family in which the inferior ovary has been shown to be of receptacular nature.

In a few genera the tip of the receptacle is sunken and the carpels line the cavity, as in *Rosa* and *Calycanthus*, but the sides of the receptacular cup are not fused to the carpels and the ovaries remain superior. The fleshy fruit of the rose is, in its lower part, receptacle; in its upper part, it consists of fused appendages, as clearly shown by the course of the vascular bundles. Stages in the development of the peculiar rose-type of fruit are shown by related genera. It is commonly believed that the rose hip and the apple represent fleshy receptacles of similar nature. But the rose and the apple tree belong to widely separated groups within the rose family, as shown by taxonomic and cytological, as well as anatomical evidence, and their fruits are not homologous. The ovary and fruit of the apple tree and closely related genera are entirely appendicular–the traces to all the floral parts arise from a receptacle of normal form and there are no inverted stelar bundles. Adnation is extensive: in one set of radii, sepal median trace, stamen trace, and dorsal-carpellary trace are united; in radii alternate with these, the lateral traces of adjacent sepals, petal trace, and three stamen traces are united.

Fusion of the traces of the various organs extends for different distances, that of the dorsal carpellary with other traces only a short distance. The ventral carpellary traces alone show no fusion with other traces. Histologically the limits of the carpels can be seen in some varieties of apple. Morphologically, the flesh of the apple consists of tissues of all parts of the flower, with the receptacle forming only an insignificant bit at the base. Genera related to the apple show stages in the evolutionary development of the ovary. The inferior ovary in most families–orchids, irises, evening primroses, for example–shows extreme fusion of the bundles of all appendages so that a cross section shows only a few bundles in the outer ovary wall.

Adnation of Flowers with Other Flowers and Other Organs

Two or more flowers borne close together on a peduncle may become anatomically fused, as in *Mitchella*, *Maclura*, some species of *Lonicera* and *Cornus*. Such fusion usually involves only the basal parts, especially the ovaries. Various species of *Lonicera* show all stages in the fusion of the ovaries of two flowers; *Mitchella* shows complete fusion of the ovaries of two, rarely three and four flowers. In these genera, fusion is so complete in the most intimately fused species that some of the bundles of one flower are fused to those of the other. One bundle within the double ovary may supply perianth parts of two flowers. Bracts may be fused to the flowers that they subtend, as in *Juglans*, and *Lonicera*. Fusion of all the parts of an inflorescence–flowers, bracts, and axis–is illustrated by the pineapple (*Ananas comosus*). The many flowers, each subtended by a bract, are closely packed in spirals.

The inferior ovary of each flower is completely sheathed by bracts, the anterior half by its own bract, the remainder by parts of three other bracts. Fusion between the ovary and the four surrounding bracts is histologically complete to the top of the ovary. This fusion extends throughout the inflorescence, uniting all ovaries, all bracts, and the axis into one structure. The bundles of the flower are wholly free from each other and from those of the bracts, but bundles of adjacent bracts are often fused and supply two bracts. Where vegetative parts are similarly fused, for example, bract to peduncle (*Tilia*); axillary peduncles to the adjacent stem (*Sparganium*, *Streptopus*)–giving supra-axillary inflorescences-the vascular tissue may also be fused and structurally complex.

Placental Vascular Supply

Because in many flowers the ventral bundles give rise directly to ovular traces, these bundles have been, incorrectly, called the "placental supply." In follicles, achenes, and some other types of ovaries, the placenta is merely a position, and there is no placental supply as such. Where the placenta is a definite, often fleshy enlargement of an area along the margin of the carpel, it is supplied by branches from the ventrals, which branching further, give rise to the ovule traces, as in the Ericaceae and Cucurbitaceae. The ovules on a placenta in a syncarpous ovary obviously may receive their vascular supply from different carpels. Where the number of ovules has been reduced from an earlier larger number, the surviving ovules may have strong vascular traces; in extreme reduction–some types of basal and free-central placentation–one or two ovules may have the vascular supply of more than one carpel (Juglandaceae; Polygonaceae). (Anatomical structure demonstrates that no angiosperm ovules are cauline.)

Vestigial Vascular Tissue

Evidence of the structure of ancestral floral types is frequently found in the vascular skeleton of modern flowers. The presence of vestiges of the vascular stele at the tip of the receptacle has been discussed. Remnants of the vascular supply of lost organs may be present in the tissues of the receptacle when all external evidence of the organs has disappeared. Only rarely does the vascular supply of a vestigial organ disappear while external remnants are still present. The vascular vestiges of lost appendages are usually stubs of traces buried in the cortex of the receptacle. The lost organs so represented may be entire whorls, especially of petals and stamens, or individual members of persisting whorls.

Traces to lost petals are common in apetalous flowers: *Aristolochia*; *Rhamnus* (apetalous forms); *Salix*; *Quercus* (some species); to lost stamens, for example, in Scrophulariaceae, Labiatae, Cucurbitaceae, *Lysimachia*, *Chionanthus*, *Lychnis*, to lost carpels in Caprifoliaceae, Ericaceae, Rutaceae, Valerianaceae. In unisexual flowers there are commonly external remnants of lost organs; in those in which such vestiges have disappeared, vascular traces may be present within the receptacle: Fagaceae; Caryophyllaceae; Urticaceae. Traces to lost ovules are among the commonest vestigial structures in the flower. They are frequent in some families, especially the Ranunculaceae, Umbelliferae, and Rosaceae.

Reduction of Vascular Supply within an Organ

As reduction occurs in an organ in evolutionary specialization–for example, as a follicle with several ovules becomes an achene with one ovule; as a corolla becomes greatly reduced in size in the crowding together of small flowers in a head, as in the Compositae–the vascular supply of the organ is greatly reduced. In carpels either the dorsal or the ventral bundles may be shortened, the latter extending only to the ovules. In extreme reduction the entire carpellary supply consists of a small bundle running directly from the receptacular stele to the ovule. Stages and variety in this loss are shown in achenes. In the stamen, the bundle, which commonly extends to the anther, may enter only the base of the filament. In the petal, either median or lateral bundles may be shortened or lost. In the gamopetalous corolla, the median bundle is more often lost and the fused laterals persists; the latter may be reduced in length.

The Sepal and the Petal

Both the sepal and the petal are commonly leaf-like in form and general appearance. Structurally also the former is like the leaf except when it is petaloid. The petal is leaf-like in grosser structure but histologically differs in many ways from the typical leaf. The vascular system is commonly reduced in amount and in supporting tissues. The mesophyll is simple in structure : there is usually no palisade layer and the spongy tissue consists of few cell layers. Chromoplasts or coloured cell sap, or both, are present. The epidermis is less simple than that of leaves. Its cells are commonly weak-walled and of complex form, often undulate, stellate, or irregularly lobed in outline and dovetailed with one another. With this shape and arrangement they doubtless form a layer mechanically stronger than one consisting of cells of simpler form. The cells of one or both surfaces are usually papillose. Stomata are fewer than those on leaves and frequently nonfunctional or absent. The guard cells lack chloroplasts except when the mesophyll also has these plastids.

Secretory areas and glandular hairs are common. Intercellular spaces, usually lacking in the epidermis of leaves, are frequent. They lie in the loops or lobes of the cell wall but are always covered by cuticle. The cuticle varies greatly in thickness being very thin on delicate, ephemeral petals. The extreme types of petals show many variations from, and exceptions to, the typical structure described above. The less highly specialized type of petal has a strong vascular supply, a weak palisade layer, chloroplasts, a nonpapillose epidermis with many stomata and often a hypodermis, and supporting tissue about the larger veins.

The most highly specialized type has the vascular tissue weak, with veinlets and even most of the main bundles lost, and wholly without accompanying supporting tissue. The mesophyll consists of one to three poorly defined layers of widely spaced cells which are lacking along the margins and in distal parts, so that part of the petal consists of epidermal layers only. The epidermis is papillose, with chromoplasts or coloured cell sap, and without stomata or glands.

The Stamen and the Carpel

Stamens and carpels consist chiefly of unspecialized parenchyma. Their trace continue in the organs as weak or strong, often amphicribral bundles, for varying distances, and with varying amounts of branching. In the stamen, the trace or traces commonly run unbranched to the anther, but in delicate stamens may fade out anywhere in the filament. In some stout stamens, as in those of *Magnolia*, small lateral branches are present. The traces of the carpel may pass as unbranched bundles to the style or stigma, or may branch in varying degrees, some showing leaf-like venation in branch. In reduced carpels, all branching is lost, and the main bundles are shortened by loss of the distal portions. For example, the ventral bundles may not extend beyond the placentae and all bundles may fail to enter the style.

In extreme reduction, as in small achenes, the vascular supply may be reduced to little more than a basal remnant. Vestigial bundles are frequently found. These are the weak or abortive remnants of traces or bundles representing the supply to organs lost in evolutionary modifications. The most common vestigial bundles are the traces to lost petals, stamens, and carpels, represented by stubs arising from the receptacular stele as normal traces but ending blindly in the cortex of the receptacle, and ovule traces that end blindly in the placenta or the margin of the carpel. The epidermis is of typical structure, the cells usually with straight walls and thin cuticle. Stomata occur on stamens only when these structures are expanded and leaf-like. In carpels they occur freely on the outer surface and are even found in the epidermis lining the ovarian cavity.

The Inflorescence

The inflorescence is the part of a plant which is concerned with the production of flowers. It may or may not be distinct from the vegetative part. Thus, in some annual plants the entire branch-system of the shoot may be regarded as an inflorescence, whilst in biennials the second year's growth may be so regarded. Inflorescences are classified according to the type of branching into (a) indefinite or racemose and (b) definite or cymose. The leaves, whether ordinary vegetative leaves or modified ones which subtend the flowering branches are termed bracts. In racemose inflorescences the growing point of the primary axis does not terminate in a flower, but, throughout its growing period, it is continually developing the axis and new lateral members, *i.e.* it is a monopodium, whereas in cymose inflorescences (sympodia) the primary axis and the successive daughter axes terminate in flowers. It is of course, possible for the primary axis to be monopodial, and the lateral branch-systems to be sympodial, thus forming a compound racemocymose inflorescence.

Simple Racemose Inflorescences

Of these we recognize four chief types:

(a) The typical raceme, in which the mother-axis (peduncle) is elongated, and the flowers are stalked (pedicellate). Examples are found in *Crotalaria* and many orchids. Similar to this in essential characters is the corymb, which may be regarded as a modification of the typical raceme. The mother-axis is relatively shorter, and, owing to the elongation of the lower pedicels, all the flowers come to one level. This renders the whole inflorescence more conspicuous to insects, and the individual flowers tend to remain small. Good examples are found in many Cruciferae (*e.g.* candytuft). Inflorescences intermediate in character between the corymb and typical raceme are described as corymbose racemes, *e.g.* the

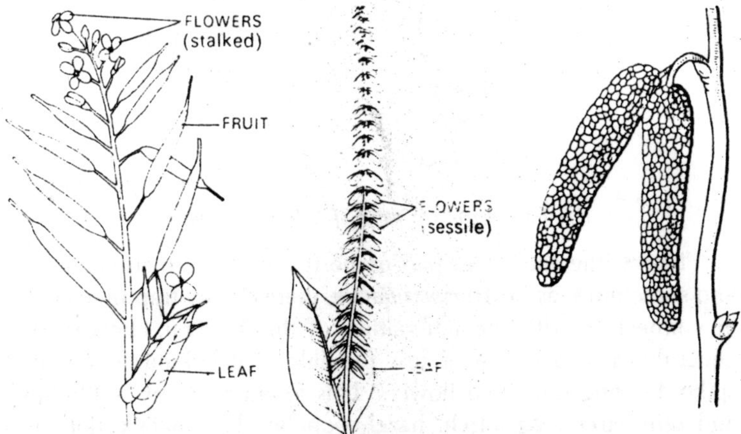

Fig. 1.34. A–Raceme of mustard, B–Spike of Achyranthes aspara, C–Catkin of mulberry.

wallflower, in which the inflorescence is corymbose when young, but lengthens out when fruiting.

(b) The spike is a racemose inflorescence in which the mother-axis is elongated, and the flowers are *sessile*, *e.g.* spotted orchid and *Plantago*. By this arrangement small flowers may be aggregated in a cylindrical mass. There are one or two special forms of the spike. The spadix is a massive fleshy spike, bearing small, usually unisexual flowers occurring especially in the Araceae. It is protected by a large enveloping bract, sometimes green more usually petaloid, known as a *spathe*.

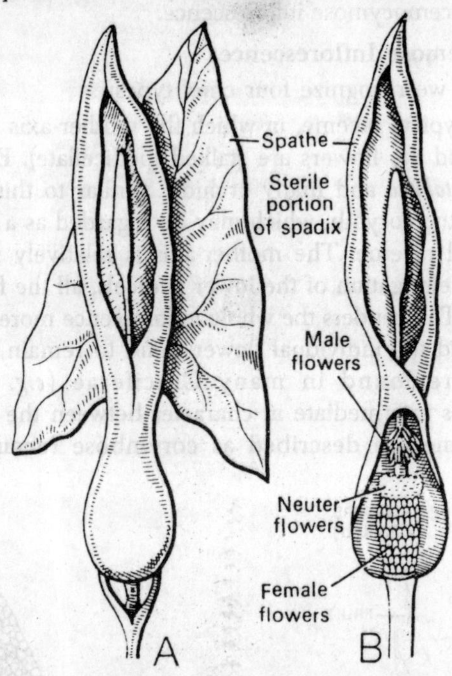

Fig. 1.35. A and B–Spadix of Colocasia antiquorum.

The spathe and upper part of the spadix serve to attract insects, and sometimes, as in *Arummaculatum*, there is a fly-trap mechanism in connection with the pollination of the flowers. The catkin is generally a long, more or less pendulous, deciduous compound spike, bearing unisexual flowers. It is found in many nut-bearing and other trees, *e.g.* birch, hazel, popular. The male catkin as a rule hangs loosely in the air so that the microspores (pollen),

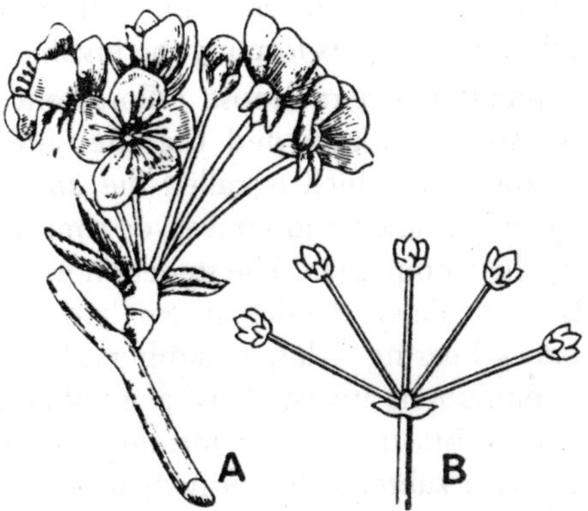

Fig. 1.36. A–Umbel of Prunnus cerasus (young), B–Outline plan.

protected from rain by the catkin scales, are readily blown out by the wind when dry.

(c) The umbel is a racemose inflorescence in which the flowers are stalked, but, owing to the abbreviation of the mother-axis, are given off at one level. The indefinite growing point develops a large number of lateral flowers, but does not give rise to an

(young), B. Outline plan.

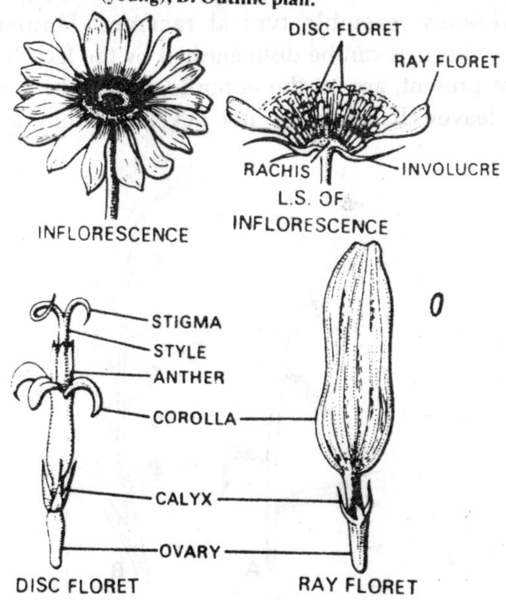

Fig. 1.37. Capitulum of sunflower with disc and ray florets.

elongated mother-axis. Each flower of umbel may be subtended by a bract, the collection of bracts forming the involucre of the umbel.

(d) The capitulum is a racemose inflorescence in which the flowers are sessile, and crowded together on a reduced or abbreviated mother-axis which is called the *disc.* It is sometimes flat, more frequently dilated and convex or conical. Examples are found chiefly in the Compositae. The student must clearly recognize that the heads of the daisy, dandelion, etc., are not single flowers, but inflorescences bearing a large number of sessile flowers or florets. The capitulum is invested by a number of small scaly, overlapping, barren bracts, together forming the involucre. Bracts (or paleae) may also subtend the individual florets as in *Anthemis nobilis* (chamomile). The massing together of small flowers in the umbel and capitulum has the same biological significance as in the corymb.

Cymose Inflorescences

These are usually either *uniparous* or *biparous.* In uniparous (or monochasial) forms each successive axis ends in a flower after producing one daughter-axis. Four different types are recognized : the bostryx, drepanium, cincinnus and rhipidium. They are *sympodial,* and sometimes resemble typical racemes. Uniparous cymes resembling racemes can be distinguished by the fact that the bracts, if they are present, are on the opposite side of the sympodial axis from the leaves. If bracts are not developed they are not easily

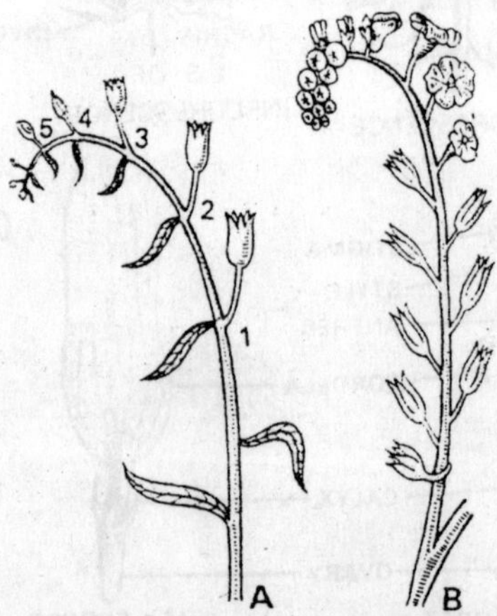

Fig. 1.38. A–Helicoid cyme outline plan, B–Myosotis palustris.

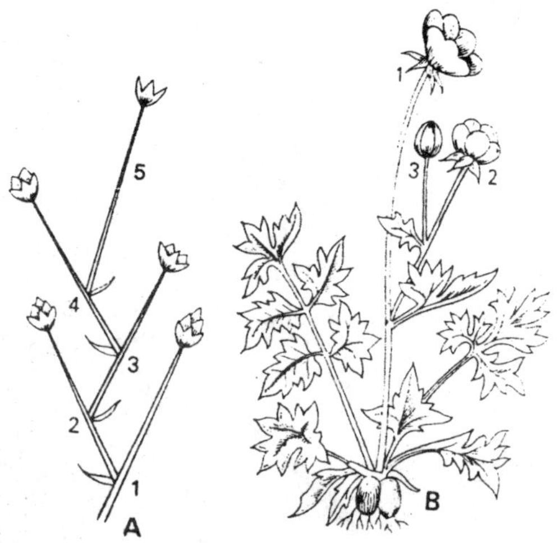

Fig. 1.39. A–Scorpoid cyme outline plan, B–Ranunculus bulbosus.

distinguished. In *biparous* cymes each axis ends in a flower after producing two daughter-axes. It is also called a dichasium. Typical examples are found in many Caryophyllaceae. Sometimes the daughter-axes are not given off at the same level, *e.g.* some *Ranunculus*, *Helleborus*, etc.

Compound and Mixed Inflorescences

Many inflorescences are compound, *e.g.* a raceme of racemes, (e.g. *Delphinium*), or spikelets (e.g. *Triticum*), or an umbel of umbels (*e.g.* most Umbelliferae). The *compound umbel*, in addition to the bracts at the base of the chief branches constituting the *involucre*, may have smaller bracts at the base of each secondary umbel the *involucel*.

Many inflorescences are mixed. We may, for example, have a raceme of spikes, a raceme of capitula, a spike of capitula, a raceme of cymes, etc. The raceme of *spikelets*, i.e the *panicle*, is a common form in many grasses (*e.g.* oat); in *Aesculus* there is a raceme of short cymes (cincinni). In *Syringa* the inflorescence is of the same nature, but the branching is much more copious.

Special Forms

There are many inflorescences which, owing to abbreviation of axes or special crowding of the flowers, do not so readily yield to careful analysis. In *Crataegus*, for example, the inflorescence might be mistaken for a typical corymb. Examination will show, however,

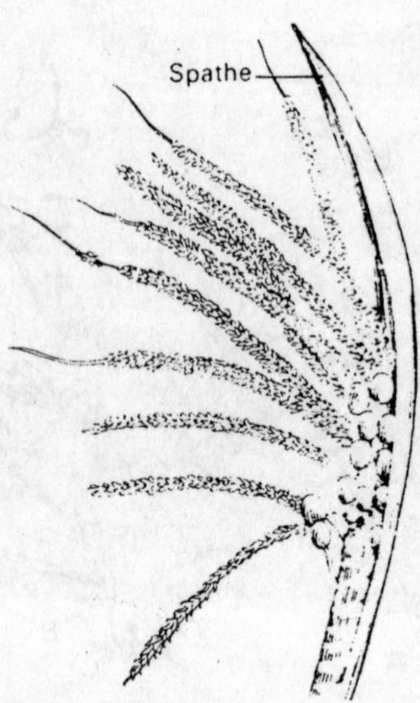

Fig. 1.40. Compound spadix (Coconut).

that the lateral axes borne on the main axis are really cymes. It is a *corymbose cyme.* Similarly the globular inflorescences of *Hydrangea* and of the snow-ball tree (*Viburnum opulus*) are corymbose cymes, rendered conspicuous by the petaloid calyces and enlarged corollas of flowers specialized for this purpose. In horticultural varieties all the flowers are sterile. In the cultivated geranium (*Pelargonium*) and many species of *Narcissus* the inflorescence, at first sight, appears

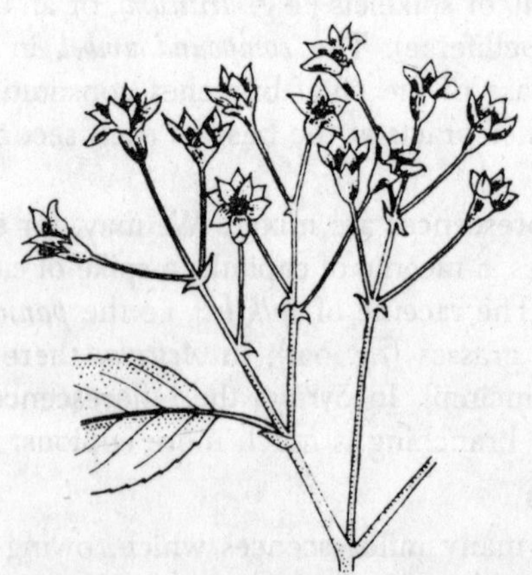

Fig. 1.41. Compound corymb (Pyrus torminalis)

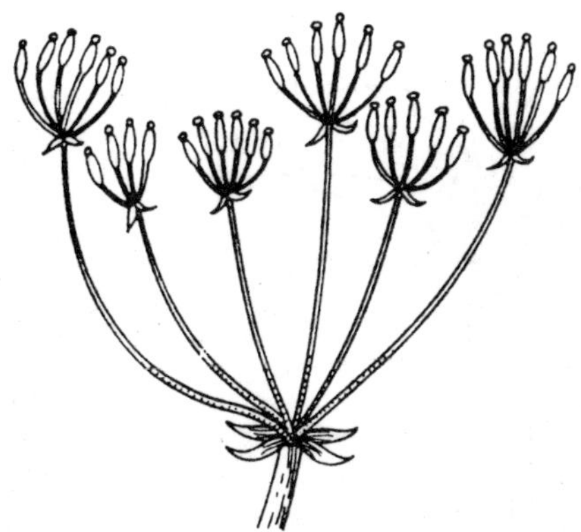

Fig. 1.42. Compound umbel (Chaerophyllum temulum).

to be an umbel. But it will be found that the young flowers are not by any means aggregated towards the centre, and that the flowers are arranged in a number of groups. These are really cymose clusters. We may speak of the whole inflorescence as an *umbellate cymose head.* These are found in many plants. In *Narcissus* the inflorescence is protected by a membranous *spathe.*

In deadnettle (*Lamium*) and many other members of the Labiatae, the leaves are opposite and decussate, and at each node there seems to be a whorl of flowers. These *apparent* whorls are called verticillasters. Careful analysis shows that there is in the axil of each leaf an inflorescence which is a dichasium of cincinni, *i.e.* a biparous cyme which passes on either side into a uniparous form by suppression of one of the branches at each branching. It is difficult to recognize this because the axes have been reduced and the flowers are sessile. It is easily recognized in many Labiatae where the flowers have short stalks.

In *Dianthus barbatus* (Caryophyllaceae) and some other plants there is a copiously branched biparous cyme, in which the axes are short and all the flowers crowded together. The cyathium is a peculiar inflorescence found in *Euphorbia* (spurge). There is a cup-shaped involucre, the margin of which bears a number of crescent-shaped glandular scales. Inside the cup there are several stamens; also a gynaeceum borne on a stalk. Careful examination shows that each stamen is really a male or staminate flower. This is borne out by the fact that each stamen is articulated to a stalk and has a scaly bract at its base. The gynaeceum with its stalk is a female

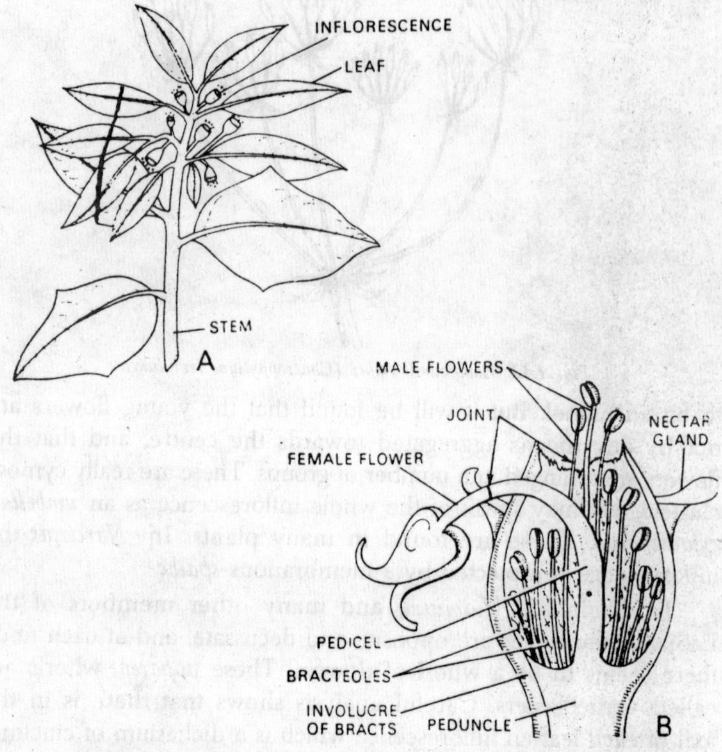

Fig. 1.43. Cyathium inflorescence. A–Twig with inflorescence, B–L.S. of one inflorescence showing a central female flower surrounded by large number of male flowers.

flower. The stamens are arranged in five groups, which are reduced cincinni, around the female flower. Thus an inflorescence looks like a single flower.

2

POLLINATION

The fertilization of the egg and the subsequent development of the seed can occur only if pollen grains have previously been deposited upon the stigma. *Pollination*, the transfer of pollen from the anther to the stigma, should be distinguished from fertilization, which is the fusion of the male and female gametes. Pollination is of two

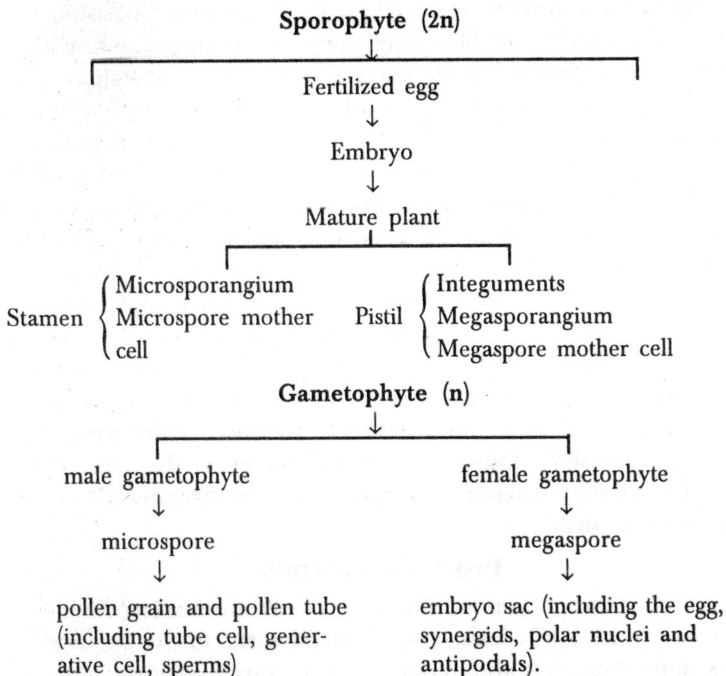

types. *Self-pollination* is the pollination of a stigma by pollen from the same flower or from another flower of the same plant. *Cross-pollination* involves the transfer of pollen from the anther of a flower of one plant to the stigma of a flower of another plant of the same or related species. The chief agents of pollination are wind and insects, but birds, snails, other small animals, and even water may carry pollen from one flower to another.

Self-pollination occurs in many plants. Among the economically important self-pollinated plants are oats, wheat, barley and rice, peas and beans, soybean, flax, cotton, and tobacco. Cross-pollination, however, is more common than self-pollination, and it probable occurs, at least occasionally, in most of the self-pollinated species.

Cross-pollination brings about a more diverse combination of hereditary units of the two parents. This results in increased variability in the offspring and greater adaptability to new environments–conditions of evolutionary advantage to the species. A more immediate effect of cross-pollination in many species is the production of more seeds or greater vigor in the offspring. A very large number of flowering plants have adaptations which prevent or reduce self-pollination. These many adaptations are usually related to pollination by insects. Prominent among them are modifications of the flower which make cross-pollination by insects possible and in some cases essential if seeds are to be formed.

So numerous and varied are these adaptations to cross-pollination that ignorance of them has caused serious difficulties when new plants have been introduced. An example of this is afforded by the history of Smyrna fig production in California. Large numbers of cuttings were introduced from Asia Minor about 1880. These grew well, but the fruits dropped from the tree before maturity. The realization that pollination is brought about by a tiny was resulted in the introduction of the insect after which the industry flourished. Information on pollination is also needed by the plant breeder when he attempts to produce varieties of greater usefulness to man.

Insect Pollination

A large proportion of flowering plants are insect pollinated. The insects are chiefly bees, wasps, butterflies, and moths, but beetles, flies, and other kinds also frequent flowers. Insect-pollinated flowers are usually brightly coloured or scented, sometimes both. Their pollen is heavy or sticky and is not readily carried by wind. Many

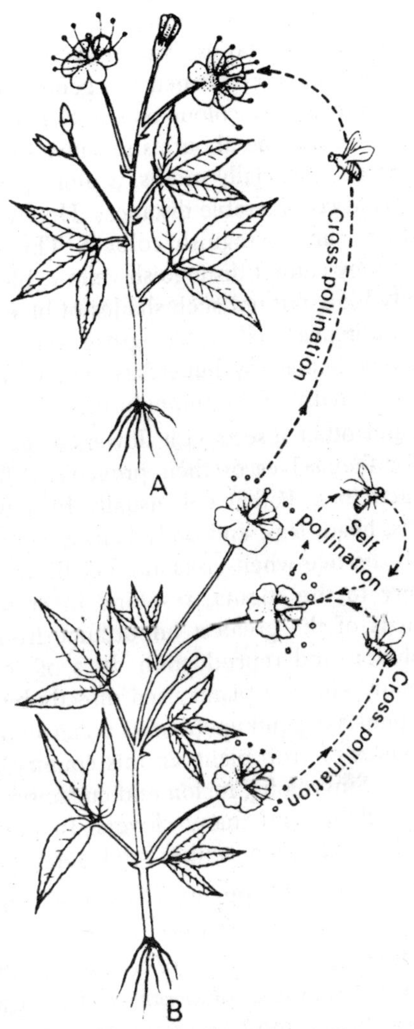

Fig. 2.1. Summary of self pollination and cross pollination.

such flowers are provided with *nectaries*, specialized tissues or organs which secrete nectar, a fluid with a sugar content varying from 4 to 65 per cent. This nectar is the raw material from which honey is made. Nectaries vary greatly in form and location and may be associated with any of the floral organs. Nectar may be secreted by the surface of the receptacle, by the walls of spurlike projections of the perianth, or by hairs on the petals or ovary.

In many flowers the nectary consists of a ring around the base of the ovary. In some, the stamens or petals have become reduced and modified into nectaries. Numerous experiments, especially on the honey bee, have confirmed the general belief that insects are attracted to flowers by colour or scent. The sense of smell of the honey bee seems to be about as sensitive as that of man, but that of other insects, especially moths, is more powerful, and they may detect odors at considerable distances. Honey bees, although partly colour blind, can distinguish colours. They are colour blind to red, which they cannot distinguish yellow, blue or dark grey. They can identify blue, but not such shades of blue as purple and violet. They also distinguish yellow, blue-green, and ultraviolet. Red flowers are usually pollinated by butterflies, which are not colourblind to red, or occasionally by hummingbirds.

Bees and other insects visit flowers to gather pollen or nectar as food for themselves or their progeny. Pollination is incidental to these activities. Pollen will usually be found adhering to the mouthparts, head, legs, and body hairs of a bee after it has visited a flower. If the bee when visits another flower, some of the pollen may adhere to the stigma, resulting in cross-pollination. Insect pollination is of the greatest importance from the standpoint of plant evolution and reproduction. It is of economic importance also, for many kinds of plants used directly by man or as food for domestic animals depend upon insect pollination. They include more than fifty kinds of crop plants in this country, among them apple and pear, the Smyrna fig, melon and cucumber, avocado, cabbage, buckwheat, alfalfa, and many clovers. In the absence of certain insects, such plants yield neither seed nor fruit.

Floral Adaptations Favouring Cross-pollination

In many plants with perfect flowers, the stamens and pistils mature at different times. This, of course, favours cross-pollination. Such a flower is said to be *dichogamous.* Dichogamy manifests itself in two ways. In the most common, the anthers ripen before the stigma matures before the pollen sacs of the stamens open. The anthers of Jacob's-ladder (*Polemonium caeruleum*), a garden ornamental, shed their pollen before the stigma of the same flower is ready to receive it; the three lobes of the stigma are folded together at this stage. After the pollen is shed, the lobes spread apart and may be pollinated by pollen from younger flowers. Flowers with pistils and stamens maturing at different times are

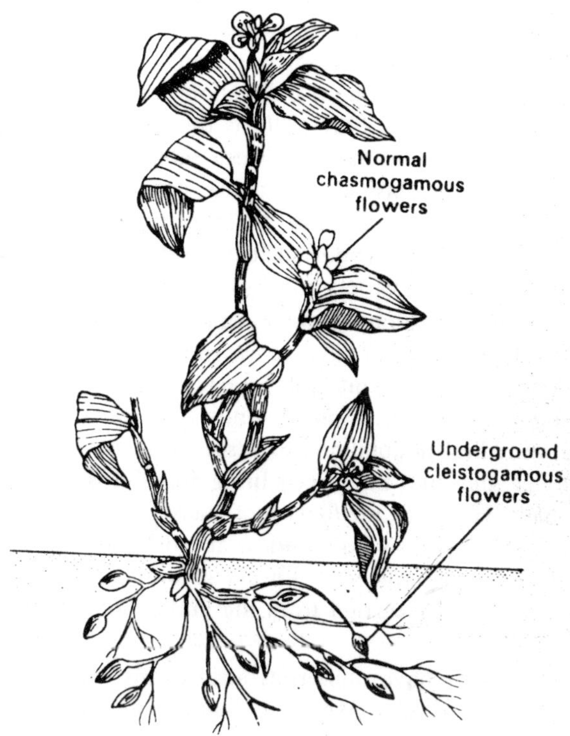

Fig. 2.2. Cleistogamous flowers of Commelina.

extremely common, and most families of flowering plants have some species in which this condition is present. The interval between the maturation of the stamens and the time when the stigma is receptive varies from an hour or so to several days. The most complex example of dichogamy is found in the avocado, widely used as a food plant in the tropics and cultivated in the warmer parts of the United States.

The approximately one hundred varieties of the avocado fall into two groups, termed class A and class B varieties. The flowers of class A varieties are functionally female in the morning and male in the afternoon. That is, in the morning the stigma is receptive but no pollen is shed, whereas in the afternoon the stigma has withered and the anthers open to discharge pollen. The flowers of class B varieties are male in the morning and female in the afternoon. During the morning, therefore, the stigmas of class A flowers are pollinated by insects which carry pollen from class B flowers. The reverse is true in the afternoon, when the stigmas of class B varieties are pollinated by pollen from anthers of class A varieties. This information on pollination in the avocado, obtained by several botanists in 1921-1926, is the basis of the recommendation that a grove of avocado trees should contain varieties of both class A and class B, to ensure cross-pollination. This is especially necessary in Florida. In other regions, irregularities in the time of blooming render such interplanting unnecessary, and the trees of a single variety may set fruit.

Even self-pollination, however, can be brought about only with the aid of bees, for the pollen of avocado is heavy and sticky and adapted to dispersal by insects. Insects which rove or climb over the floral parts pollinate many kinds of flowers which have no special structural adaptations to insect visitors. These visits take place by chance, and the insects bring about cross-pollination only occasionally. However, the flowers of many plants are so constructed that they are visited by one kind or at most a few kinds of insects. These insects regularly visit flowers of the same species, so that pollen is frequently carried from the anthers of one flower to the stigma of another.

Such flowers have numerous adaptations which favour cross-pollination and which make self pollination unlikely or impossible. The stigma is commonly placed so that it is brushed by an insect visiting the flower by does not receive pollen from the anthers of the same flower. This condition commonly results when the elongated style projects well above the anthers. In many flowers there is a close relationship between the depth of the corolla tube and the length of the sucking mouth parts of the insects which usually visit them. The nectar at the base of the corolla tube may be so deep that only a butterfly or moth with long tubular mouth parts can reach it. This is the case with a number of strongly scented white or nearly white flowers which open in the evening. Soapwort (*saponaria*), some species of tobacco, and the Jimson weed (*Datura*), which has a corolla tube 3 inches long, are pollinated only by night-flying hawkmoths.

The red clover is pollinated only by bees whose mouth parts are of sufficient length to reach the nectary at the back of the flower. Flowers with a short corolla tube or exposed nectar are likely to be visited by a variety of insects, including flies. Cross-pollination is also promoted in plants in which the length of the

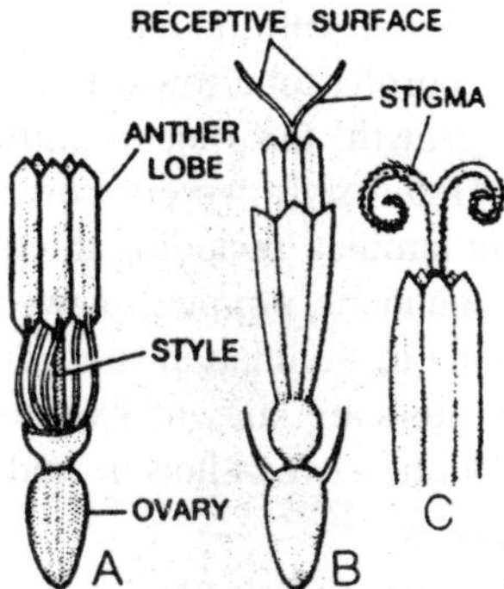

Fig. 2.3. A–Protandrous flower of Sunflower with mature anthers but immature stigma, hidden within the anther tube, B–Maturing and expanding bifid stigma showing receptive surface, C–Curving back of stigmatic lobes if cross pollination fails.

style differs in plants of the same species (*heterostyly*). In one kind of plant all the flowers have short styles and the anthers are above the stigma. In the other kind, the flowers all have long styles, and the anthers are located below the stigma. The mouth parts of an insect, thrust into a short-styled flower, come in contact with the anthers and become covered with pollen. If the insect should now enter a long styled flower, the pollen is deposited upon the stigma, which is located at approximately the same level as the anthers in the short-styled flower. Similarly, the pollen adhering to an insect which has visited a long-styled flower will be deposited upon the stigma of a short-styled flower, which is at the same level in the corolla as the anthers of the long-styled flower.

Self-pollination is not precluded by this structure, for pollen may fall from the anthers of a short-styled flower onto the stigma.

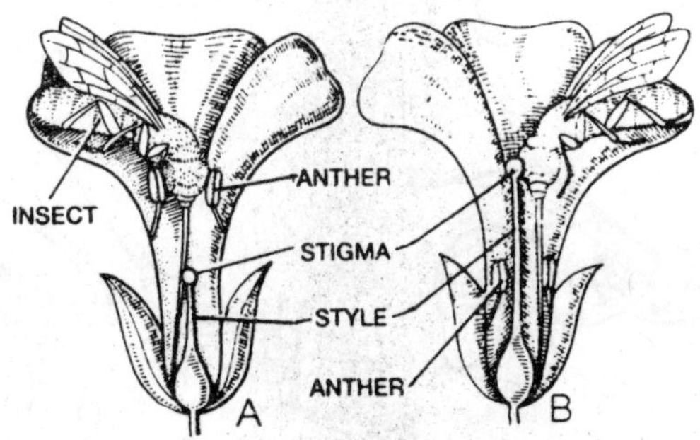

Fig. 2.4. Primula. A–Short style and highly placed anthers, B–Long style and lowered anthers.

In the long-styled flowers, self-pollination may occur because of the elongation of the corolla tube toward the end of the flowering period. In this final growth, the ring of anthers is brought to the level of the stigma. Heterostyly is very common. It is found in a large number of plant families, including the pinks, St.-John's-worts, heaths, wood sorrels, gentians, poppies, pigweeds, and many others. Innumerable variations in pollination mechanisms and adaptation of flowers to insect visitors are known. Two final examples illustrate how complex the relation between flowers and insects have become.

Pollination in Salvia

The salvias (sages) have a pollination mechanism whereby a sudden movement of the filaments results in the transfer of pollen to an insect visitor. This mechanism is best known in *Salvia pratensis*, a cultivated ornamental herb with blue flowers. The flower is irregular, with a strongly developed upper and lower lip. Nectar is secreted by a ring of glands by the base of the ovary. Only two stamens are found, concealed under the upper lip.

Each filament is jointed near the base and is prolonged downward into a shell-shaped expansion which connects with a similar expansion from the other filament to form a plate which closes the mouth of the corolla tube. When a bee pushes into the

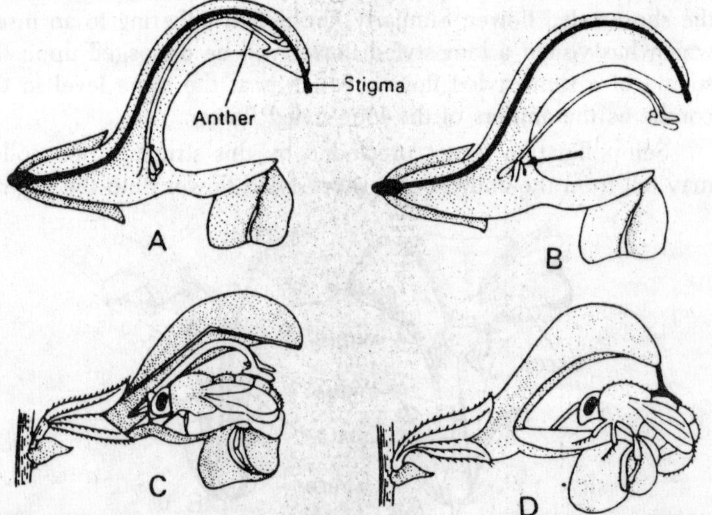

Fig. 2.5. Pollination in Salvia. A, B–L.S. of flower showing immature pistil and mature stamens, C–A bee coming with pollen, D–A flower with mature stigma.

flower to obtain nectar, its mouth parts encounter this plate, which is then pushed upward and backward, causing the filaments and their anthers to tip forward. The anthers then come in contact with the back of the bee, which is dusted with pollen. When the insect withdraws, the filaments return to their former position. The stigma, located above the anthers, now matures; it grows downward and the lobes open. If the bee then visits an older flower, it brushes against the downwardly directed lobes of the stigma, to which the pollen is transferred. Pollination of the stigma by pollen from the same flower is, of course, impossible.

Pollination in Yucca

The relationship between the Pronuba moth and the pollination of the flowers of the yucca plant is so extraordinary as to be unbelievable if it had not been verified repeatedly by qualified investigators since its discovery in 1892. The yuccas are a group of some thirty species of lilly-like plants with long, sword-like leaves. They are popularly called Spanish bayonet or Spanish dagger. Most the native to the arid parts of the Southwest and Mexico; some are cultivated as ornamentals. In the early summer the plant sends up a tall, coarse flower stalk bearing numerous large, white, drooping, bell-shaped flowers. The stamens, six in number, are much shorter than the pistil and also arch away from it, so that self-pollination is impossible. The style is short, and the

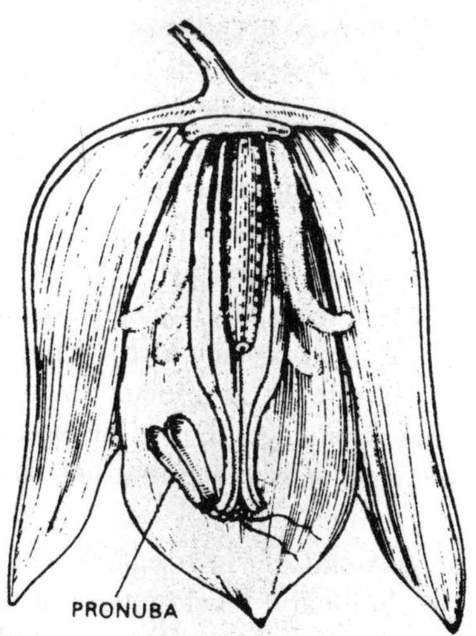

Fig. 2.6. A mature Yucca flower with Pronuba moth pollinating the flower.

stigma is three-lobed, with a deep chamber lying between the lobes. In the evening after the flowers open, the silvery white Pronuba moths, about 5/8 inch long, appear in the vicinity.

The moths may come from a distance, attracted by the wind borne fragrance. Both male and female Pronuba moths take no food and live only 2 to 5 days after mating. The males and females meet within the flower and mate. The female, then places herself at the top of a stamen, where she scrapes pollen from the anther by means of highly modified mouth parts. Visits to several stamens produce a large amount of pollen, which is kneaded into a ball held just under the moth's head. The female then flies to another plant and, after boring a hole in the soft tissues of the ovary, deposits an eggs in the vicinity of the ovules. As soon as the egg is laid, the female moves to the top of the pistil and pushes a portion of her load of pollen into the stigmatic chamber, pressing it down firmly with several vigorous thrusts. Another egg is then deposited within the same ovary, and again a portion of the pollen mass is forced into the stigmatic chamber.

The depositing of each egg is followed by pollination, until about five eggs have been laid. The egg hatches into a larva, which feeds upon the seeds which have developed from the ovules. The larva matures in about a month, bores a hole in the wall of the mature ovary (capsule), and drops to the soil, where it burrows to a depth of several inches. Here it spends the winter and spring in a tough cocoon. When the yucca flowers again, the insect appears above the ground as a winged adult and the cycle is repeated.

The pollen which the female Pronuba has placed upon the stigma serves as a supply of food for the larva when it emerges from the egg. Without pollination, the ovules would not develop into seeds with their stores of readily available food. The yucca, in turn, would be unable to reproduce without the Pronuba. Each larva eats relatively few seeds, perhaps eighteen to twenty-five, and the several hundred sound seeds which remain in each fruit allow reproduction to continue. It is difficult to avoid the conclusion that the moth behaves intelligently and purposefully in this curious relationship, perceiving in advance that her labours result in food for her young.

A more rational point of view regards this relationship as the result of a long evolutionary development, in the course of which the floral structure and the mouth parts of the insect have all become modified from a simpler condition. In whatever manner the flower and the insect became mutually adapted, it is now certain that without one there could not be the other.

Self-sterility

In many plants with perfect flowers, fertilization and seed formation follow self-pollination. This condition is known as *self-fertility.* In other plants with perfect flowers, fertilization does not occur when the stigma is pollinated by pollen of the same flower. Even pollen from another flower of the same plant or from flowers of other plants of like hereditary constitution will not result in fertilization. This condition of physiological incompatibility is known as *self-sterility.* Many species which show the phenomenon of self-sterility can be arranged in groups of individuals which fail to set seed when pollinated by other individuals of other groups. Several hundred such species, wild and cultivated, are known, and the list is probably far from complete.

Most cases of self sterility are due to the slow rate of pollen-tube growth when the flower is pollinated by pollen from flowers of like heredity. The rate of pollen-tube growth is determined by

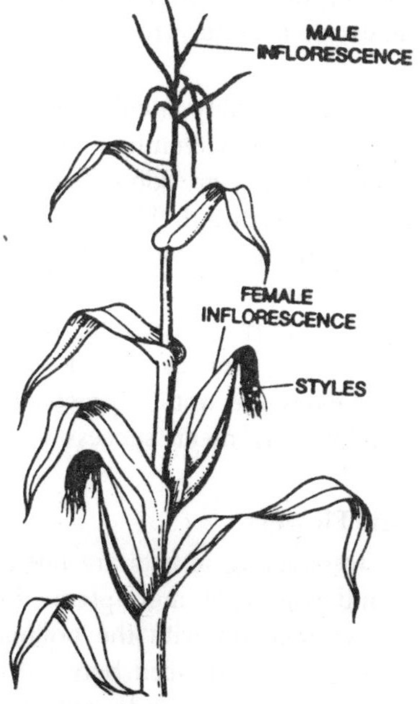

Fig. 2.7. Maize cob showing long silky style and stigma for catching pollen grains.

certain genes. If the pollen tubes carry the same sterility genes as the style, the rate of growth is so slow that the flower withers and falls away before fertilization takes place. If the genes of the style are suitably different from those of the pollen tube, the tube grows rapidly and fertilization follows. In most ornamental plants self-sterility is of little significance because these plants are grown for their flowers of foliage. But self-sterility in such plants as red clover, apples, pears, cherries, and plums, which are cultivated for their seeds or fruits, is another matter.

Some varieties of apples set fruits well if self-pollinated or pollinated by pollen from flowers of other trees of the same variety. Most apple varieties are self-sterile, and commercial orchards are therefore planted with at least two varieties whose pollen is effective in cross pollination and whose blooming periods overlap. Certain combinations of varieties are unsuccessful, and the grower must have a list of the varieties which are compatible when crossed. The varieties are set out in the orchard so that several rows of one variety alternate with rows of another. Bees bring about cross-pollination. Frequently hives are rented from beekeepers and places in the orchard during the period of bloom. Even many of the self-pollinated varieties set a better crop when crossed with another variety.

The existence of self-sterility in cultivated fruits has been known since the latter part of the nineteenth century, but even today large apple orchards of one variety are sometimes planted, with the result that little or no fruit is produced. This can be overcome, however, either by grafting other varieties to established trees or by placing bouquets of apple flowers of another variety in the orchard during the pollinating period. Pears and plums, like apples, include some self-sterile varieties. Sweet cherries are self-stertile, but sour cherries, nectarines, and peaches are not. A practical knowledge of these aspects of pollination may mean success or failure to the grower of fruit trees.

Insects that Visit Flowers

The chief flower-visiting insects are beetles (Coleoptera), flies (Diptera), bees and wasps (Hymenoptera), butterflies and moths (Lepidoptera). In connection with the pollination of flowers, the important differences to be noticed between these insects are the size of the body, the length of the tongue (proboscis), the time of year at which each kind is most plentiful, and their habits–*e.g.*, whether they collect pollen or nector or both, whether they fly by day or in the evening. By carefully studying the structure of a flower, and nothing such points as the time of flowering, the order in which the anthers and stigmas mature, the relative positions of anthers and stigmas in the open flower and any changes in position that may occur, we can often tell what kind of insect is capable of effecting cross-pollination, and whether or not self-pollination is possible. Most flies and beetles have very short tongues, usually less than 3 mm. long. Most of the larger and longer-tongued flies, *e.g.*, gadflies, "cleggs," and horseflies do not visit flowers; but there are some, chiefly hover-lies and bee-flies, with tongues sometimes as long as 12 mm., which are regular flower-visitors. Flowers may be arranged in various biological group or classes according to their adaptations for insect-visitation:

Flowers adapted for short-tongued insects

These may be (*a*) flowers in which the necter is freely exposed on the surface, *e.g.*, Umbelliferae, Rutaceae, etc. : (*b*) flowers with a very short tube, *e.g.*, moschatel, bedstraw, enchanter's nightshade; (*c*) shallow open flowers such as stonecrop and saxifrages. Such flowers are visited by the shorter-tongued beltles and flies.

Flowers with partially-concealed nectar

This group includes flowers in which the nectar can be reached only by insects with tongues at least 3 mm. in length, and which are therefore visited by the longer-tongued beetles and flies, as well as by insects of higher type. The nectar may be slightly concealed by the stamens, *e.g.*, *Ranunculus* and Myrtaceae; by the erect stiff sepals, as in the smaller Cruciferae; by the formation of a shallow calyx-tube, as in many Rosaceae; by a short corolla tube, *e.g.*, the shorter-tubed Compositae, etc.

Flowers with fully-concealed nectar

This type of flower differs only in degree from the last. Here the nectar can only be reached by insects having tongues about 6 mm. long, including the longest-tongued flies (chiefly hover-flies), the shorter-tongued bees, and wasps. The concealment of the nectar is effected by a further deepening of the flower, owing to the formation of a calyx-tube, to the calyx being gamosepalous or the corolla gamopetalous, or to other causes. Examples of these medium-tubed flowers are seen in some Rosaceae, Solanaceae, Scrophulariaceae, etc.

Long-tubed flowers

When the flower-tube becomes longer, all the shorter-tongued insects are more or less completely excluded, and the flower is adapted for, and chiefly visited by, the larger bees, butterflies and moths. Many flowers belonging to the Amaryllidaceae, Iridaceae, Orchidaceae and Scitamineae, in which the perianth nearly always forms a long tube, come under this type. Flowers like those of Papilionateae, *Antirrhinum*, and *Calceolaria*, can only be opened by large bees, and only the longest-tongued bees can reach the nectar in such flowers as *Aconitum* and *Delphinium*.

Humble and hive-bees have the most perfect mechanism (the "pollen-baskets" on the hind-legs) for collecting pollen to mix with nectar and feed their broods. Humble-bees have longer tongues than hive-bees, and are particularly skilful in finding the way to well-concealed honey.

Blue, purple, and red colours are often associated with flowers visited by bees (especially blue and purple) and butterflies (especially red), while flowers visited by other insects are usually white, yellow, or variegated; but there are far too many exceptions to allow of a general rule.

Birds, especially small honey suckers and humming birds, may be active as pollinators. Flowers may be visited by both insects and birds, but bird-pollinated flowers generally show brilliant colour contrasts. Examples are afforded by *Strelitzia regina*, many species of *Protea*, *Salvia* species, *Loranthus aphyllus*, and some *Acacia* species.

Butterfly-and moth-flowers

When the flower-tube (or at any rate the level of the nectar) is more than about 12 mm. (about half an inch) deep, the nectar is beyond the reach of bees, though they may visit the flower for pollen, or the humble-bee may bite through the tube (calyx or corolla) and thus rob the flower of its nectar. Butterflies may visit many flowers which are adapted for bees, most butterflies and moths having tongues of about the same length as, or a little longer than, those of bees.

Some moths, however, have far longer tongues (30 mm. or more in British species), which are (an in butterflies) carried coiled up in a spiral under the head when flying. These moths can reach nectar when it is at the bottom or a very long tube, as in *Lonicera*, which is visited chiefly by the night-flying privet hawk-moth, and the white convolvulus, which is pollinated by another species of hawk-moth (*Sphinx convolvuli*, tongue 80 mm long). Other flowers pollinated by night-flying moths are *Oenothera, Nicotiana, Ligustrum* and *Cereus*. Moth-pollinated flowers are generally white or pale-coloured, sweetly scented, and open in the evening, usually remaining closed and almost scentless during the day.

Examples of Floral Mechanism

In the garden pansy (*Viola altaica*) the authers of the five stamens are firmly joined by hairs on their edges, and the two anterior stamens bear processes, functioning as nectar-glands, which pass down into the spur of the anterior petal. A space or chamber ("pollen-box") is enclosed above the ovary, at the base of the style, by the five membranous scales borne on top of the authers. The stigma, which projects beyond the auther-scales, is dilated and hollow. It has a tuft of hairs on each side, and below there is an opening into it, the lower edge of which is protected by a lip or flap (the "scraper").

The flowers are pendulous, and hence the pollen, which is shed on the inner faces of the anthers, and is dry and loose, not sticky as in most entomophilous flowers, falls into the "pollen-box," from which it can escape only through the opening between the scales of the two anterior anthers.

The flowers are pollinated by long-tongued bees (and butterflies). When the insect enters the flower, pollen obtained from another flower may be deposited on the stigma, and cross-pollination thus effected. Pushing down into the spur of the anterior petal to reach the nectar, the insect receives a supply of pollen which has escaped from the "pollen-box". The "scraper" prevents this pollen being deposited on the stigma as the insect retires. The conspicuously coloured centre of the flower, and the honey-guides on the lateral and spur petals serve to attract desirable insects. The entrance of small undesirable visitors is hindered by the hairs on the lateral petals and on the sides of the stigma, by the hairs lining the entrance and cavity of the spur, and by the length of the spur itself.

In *Salvia* (sage), one of the Labiatae, an interesting mechanism is found. The corolla is bilabiate. The conspicuous lower lip attracts insects, and acts as a landing-place. The arched upper lip protects the stamens and style. These are only two stamens, the other two, characteristic of the Labiatae, being represented in the sage by staminodes. The two stamens have a peculiar structure. Each has a short filament, jointed to a long curved connective. In some types of *Salvia* each end of the connective bears a half-anther, but in other types (*e.g.*, garden sage) the lower end of the connective is barren and flattened and the upper part of the connective is longer than the lower, the whole forming a delicate lever.

A bee on entering the flower pushes against the *united* lower ends of the two connectives in seeking the nectar, and causes the curved connectives to swing on the filaments as on hinges, so that the two fertile anther-lobes (*a*) come down and strike on the bee's back, dusting it with pollen. As the bee retires, the stamens return to their former place under the corollahood. The flowers are protandrous. As the flower gets older the style curves down, and the stigma is so placed that it is touched by a bee entering the flower at this stage.

Wind Pollination

In many ways the simplest form of pollination, and probably the most primitive, is pollination by wind. The flowers of wind-pollinated plants are usually small and inconspicuous. Devoid of bright colours and nectar, they are commonly grouped in dense clusters. The perianth is often reduced or lacking. The stamens may protrude prominently from the flower, suspended by long filaments. The stigmas are frequently large and branched. The pollen grains are small, light, and dry, and are commonly produced in larger quantities than in insect-pollinated flowers.

Wind pollination is found both in woody plants and herbs. Among wind-pollinated woody plants are the conifers, the poplars, oaks, ashes, elms, birches, hickories, and sycamore. In many woody plants the staminate and sometimes the pistillate flowers are grouped into drooping clusters called catkins. The wind-pollinated herbaceous plants include pigweeds, sorrels, docks, plantains, nettles, meadow rue, hops grasses, rushes, and sedges. Some species may be either wind or insect pollinated. The small flowers of willows, for example, are brightly coloured and provided with nectaries. But if the pollen is not removed by insects, it may be carried away by wind. Many maples, too, are both wind and insect pollinated.

Hay Fever

Large numbers of people are aware, to their sorrow, that numerous wild and cultivated plants are wind pollinated. These people are sensitive or susceptible to the pollen of such plants, a condition known are *allergy*. The same pollen is harmless to other individuals. The allergy to plant pollens is popularly called hay fever. In order to cause hay fever, the pollen of any given plant must have three important characteristics : it must be produced in considerable amounts; it must be buoyant, so that it floats readily in the air; and it must be toxic to allergic people.

Thus, although nearly all hay fever is produced by wind-pollinated plants, all pollen carried by the wind does not cause hay fever. The cattails shed large amounts of pollen, as do the conifers, yet except for a few species of conifers these plants do not cause hay fever, for their pollen is not toxic. The goldenrods, cosmos, roses, and sunflowers are popularly believed to be important sources of hay-fever pollen. The pollen is toxic to some people, but these plants are insect pollinated. Their pollen is heavy and is not readily carried by the wind. Only intimate contact, such as handling roses or golden-rod, will produced in an allergic person symptoms of the disease. Usually these plants are only minor offenders in causing hay fever.

Pollination Seasons

Whether or not a sensitized person will develop hay fever depends upon the kind and amount of pollen in the air. This in turn depends, except for such atmospheric conditions as rain and wind, upon the plant species which occur in a given region. The time of year when the symptoms of hay fever appear is related to the flowering period of these plants. Three hay-fever seasons are recognized in many parts of the United States. But there is great local variation in the distribution of the plants which cause hay fever and in the pollination periods of these plants.

Early spring (March to May)

The usual cause of hay fever in this period is tree pollens. Among the trees concerned are the box elder, the elms, poplars, birches, oaks, hickories, ashes and sycamore.

Late spring and Early summer (May to July)

Grasses are the most important hay-fever plants during this period. Among the more important species are Kentucky blue grass, red top, orchard grass, timothy, sweet vernal grass, and Bermuda grass.

Late summer and fall (August to September)

The ragweeds are outstanding pollen producers during this season. The plants of significance in hay fever, the pollination periods of such plants, and the quantity of pollen in the air can be determined by pollen surveys for a specific region. Atmospheric pollen counts are made by exposing to the air microscope slides covered with petroleum jelly. The pollen which adheres to the slide is stained and studied under a high powered microscope and it identified by surface features, which vary widely in pollen grains of different species. Newspapers sometimes publish daily pollen counts for the information of hay-fever victims.

Contrivances and Conditions Favouring Cross-pollination

There are in flowers many arrangements and mechanism which ensure cross-pollination. Usually such arrangements and mechanism merely give chances in favour of cross-pollination without precluding the possibility of self-pollination. Sometimes, however, they make self-pollination difficult, or altogether impossible.

In plants with unisexual flowers, of course, cross-pollination is absolutely necessary if seed is to be produced. We have this condition in its extreme form in dioecious plants, e.g., *Salix* (willow). There are a few plants, also, in which cross pollination must take place if seed is to be produced, because the plants are *self-sterile, i.e.*, the flower cannot be fertilized by its own microspores; this occurs in some specimens of *Passiflora* (passion-flower), or *Lobella*, and of *Abutlion*.

In some flowers, again, self-pollination may be rendered unlikely or difficult owing to the relative position of anthers and stigma. A condition of much more general occurrance is that known as Dichogamy. This is a condition in which the anthers and stigma in hermaphrodite flowers come to maturity at different times, and which, when completely developed, entirely prevents self-pollination.

There are two form of dichogamy : (*a*) protandry, in which the anthers ripen first, so that when the microspores are shed the stigma of the same flower is not ready to receive them; in their case, if the microspores are not to be wasted, they must be transferred to an older flower; (*b*) protogyny, in which the stigma ripens first; here the microspores must be transferred to a younger flower. Protandrous flowers are much more common than protogynous. Examples of the former are found in Compositae, Labiatae, Umbelliferae, Solanaceae, etc., of the latter in *Plantago*, *Luzula*, some Scrophulariaceae, Ranunculaceae, etc. Wind-pollinated flowers are more often protogynous than protandrous, but many are unisexual.

Anemophilous and entomophilous flowers have each special characters of their own, so that as a rule we can distinguish them at a glance. In *anemophilous flowers*, the microspores are usually dry and smooth, and produced in great abundance, as much must be wasted; the flowers are small and inconspicuous; there is no nectar or perfume; and frequently the stigmas are branched and feathery, to catch the microspores. In many trees which are wind-pollinated the flowers appear in spring before the leaves, so that the microspores have free access to the flowers.

In most herbaceous plants with wind-pollinated flowers, the latter are carried up on a long stem, well above the leaves, so as to expose them as freely as possible to the wind (*e.g.*, *Plantago*, Chenopodiaceae, grasses, etc.) Much greater variety of adaptation is shown by *entomophilous flowers*. As a rule they have large, conspicuous, or highly-coloured corollas, or are arranged in conspicuous inflorescences; they usually secrete nectar and give out perfume. The microspores are usually rough and sticky, and often not produced in any great abundance, as they are more sure of transference. The bright corollas, the perfume and nectar serve to attract insects which visit the flower in search of food. A nectarless but otherwise insect-attracting flower is sometimes called a "pollen-flower." Examples are found in *Papaver*, *Rosa*, *Helianthemum*, *Anemone*, *Clematis*, *Hypericum*, *Ulex*, *Cutisus*, *Ulmaria*. These flowers are visited by pollen-feeding insects.

Many entomophilous flowers are further characterized by the presence of ingenious mechanical devices, which guide and control the movements of the insect and turn them to the best account. Thus, in many cases the corolla is so constructed that the insect must alight on the flower or enter it in a special way (e.g., Labiatae, Papilionatae); the same result may be attained by the secretion of nectar into special receptacles or spurs (e.g., in *Viola*). Often the insect, on entering a flower, pushes against special processes or outgrowths which move the stamens and bring the anthers in contact with its body (e.g., *Salvia*, sage); or the stamens may be jerked, and the microspores scattered over the insect's body. In some flowers the stamens move in response to touch by a visiting insect. This is seen in *Centaurea jacea* where the filaments contract and pull the tube formed by the syngenesious anthers backwards over the style, and the microspores are thus swept out and exposed. Curvature movements of stamens in response to a shock stimulus are seen in a number of Families such as Berberidaceae, Tiliaceae, Cactaceae and Cistaceae. Also sensitive stigma may close over the deposited microspores as in Scrophulariaceae. Frequently spots or lines of a conspicuous colour are developed on the corolla; these have been called "honey-guides," as they are believed to afford insects guidance in seeking out the nectar. The general result of all these devices is that the insect receives pollen on a special part of its body, and when it enters another flower the pollen is deposited on the stigma.

In many protandrous flowers this is secured by the style bending over so that the stigma is in the position formerly occupied by the stamens. A special, but at the same time simple, arrangement for ensuring cross-pollination by insects in known as heterostyly. It is seen in primrose. Here there are two types of flower, borne on different plants. One kind (thrum-eyed) has long stamens (with anthers in the throat of the corolla tube) and a short style; the other (pin-eyed) has a long style and short stamens; thus in the two types the positions of anthers and stigma are reversed.

Pollination is effected by transference between these two forms and not between two flowers of the same form. This is the dimorphic form of heterostyly. In purple loosestrife (*Lythrum*) there are three types of flower combining two positions for the stamens and one for the stigma. Such flowers are said to be trimorphic. Trimorphism is also seen in species of *Oxalis* and dimorphism in *Turnera* and species of *Jasminum*.

Special Arrangements for Self-pollination

In studying floral mechanisms we are too apt to forget that self-pollination occurs regularly in most flowers were it is not precluded by dioecism, complete dichogamy, or self-sterility. Many annual plants are commonly self-pollinated (*e.g.*, groundsel, chick-weed). They have small flowers, often without nectar or smell, and are either homogamous, that is, their anthers and stigmas mature at the same time, or so slightly dichogamous that self-pollination is secure. Even in flowers evidently adapted for cross-pollination, if this fails, there is commonly the possibility of self-pollination. Many of them are distinctly dichogamous, but not completely so, there being usually a short period during which self-pollination becomes possible. To effect this there are sometimes special contrivances such as the curling back of the stigmas to reach the pollen on the anthers or style (*e.g.*., Compositae, Campanulaceae). A very special adaptation for self-pollination is the production of cleistogamous flowers. There are closed flowers produced later in the year by certain plants which had previously produced entomophillous flowers, *e.g., Viola odorata, Oxalis acetosella, Lamium amplexicaule* (one of the dendnettles), etc. *Commelina benghalensis* also may produce cleistogamous flowers on leafless shoots.

The cleistogamous flower is small and inconspicuous. The calyx remains closed, and the stamens and pistil are developed within it. In *Viola odorata* the self-pollinating cleistogamous flowers have five very small petals and five stamens, but in the dog violet there are only two (anterior) stamens. The anthers produce few microspores, and do not open; the microspores germinate inside the anther, and the pollen-tubes grow through the anther-wall and the style to reach the ovules. The formation of these flowers is partly dependent on shade; they are always shaded by the leaves of the plant itself. If a plant is kept in feeble light, it will usually produce only cleistogamous flowers.

The Essential Organs and Their Functions

The main function of the flower is the production of seeds, and for the setting of the seeds two distinct processes are necessary. They are the transference of the pollen-grains from the anthers to the stigma (pollination), and a fusion of a bit of the protoplasm of the pollen-grain with a portion of the protoplasm within the ovule (fertilization). We know that the parts of the flower direct concerned in these processes are its essential organs, the stamens and the pistil. For the clear understanding of these two processes, a knowledge of the essential organs is necessary. The stamens are the parts intended to produce the pollen-grains. A stamen usually consists of a filament and an anther, but it is the anther alone that is of importance. The filament is only a stalk supporting the anther, and we have many instances in which the filaments are absent, the stamens being reduced to mere anthers.

To study the structural details, the stamens of Tribulus may be sued, as it is typical of the anthers of a good number of plants. A transverse section this anther reveals two distinct lobes with two cavities in each, filled with pollen. But the lobes of an old anther present only a single cavity, as the two cavities get fused. The anther lobes are held together by the connective, which is traversed by a vascular bundle. The wall of the anther consists of two layers of cells, an outer layer or the epidermis and an inner layer in which the cells have special thickenings (and hence called fibrous cells). This differentiation in the cells of the wall of the anther is intended to facilitate its dehiscence. The outer wall of the cells of the epidermis bulges slightly outwards and there is protoplasm within these cells. Until the maturity of the anther these cells remain in a turgid condition. The pollen grains take some time to develop and until they are fully formed they need protection. So the wall of the anther remains intact and does not burst, till the pollen is ripe enough for shedding. As long as the epidermal cells are turgid they exert pressure on the fibrous cells laying below the epidermis. As the anther approaches maturity the epidermal cells begin to lose water gradually. This means the gradual lessening of the pressure on the fibrous cells and, as soon as they are freed from it, they tend to expand and assume their natural size.

The removal of compression on the fibrous cells leads to the curling of the anther wall. This, of course, causes the wall to break in some place where it is weak. The pollen-grain of *Tribulus terrestris* is round and it is only a bit of protoplasm enclosed by a cell wall in which, two distinct layers, anther cutinised and an inner cellulose layer, can be distinguished. The outer layer is thickened in a peculiar manner. The inner layer is smooth and uniformly thick. There is a great deal of variation in size and form, and the sculpturing of the outer layer in the case of the pollen-grains, according to the species of the plant. Those of very showy flowers are in many cases provided with spiny or other kinds of projections, or the outer layer may be sticky. In a few cases, we find some spots in the wall formed beforehand for the coming out of the pollen-tube. Some pollen-grains have special lids also, as in the case of Cucurbita.

The free end of the style or the stigma is the part intended for the reception of the pollen. It has to catch the pollen-grains and retain them, and also to assist them to germinate. The surface of the stigma is usually rough due either to papillae or hairs, and a sticky juice containing some sugar is also secreted by it. Both these conditions are useful in retaining the pollen-grains on the stigma. Sometimes the stigma becomes very much branched and is consequently plumose, as in the case of grasses. It is said that the pollen grains are safeguarded from the attacks of bacteria, by the stigmas secreting some substance detrimental to their growth.

A stigma is not always receptive; it becomes receptive as soon as the flower opens and generally continues to be so far sometime, this period varying with the kind of the plant. The sugary juice secreted by the stigma seems to be necessary to stimulate the pollen-grains to germinate. It is only when the stigma is receptive that we find the sugary juice, and if it is absent it is an indication that the stigma is not receptive.

The ovule which develops into seed after fertilization lies within the ovary. It does not develop into a seed unless a bit of protoplasm from the pollen-grain finds its way into the interior of the ovule and mixes with a definite part of the ovule. When the ovule is fully formed and ready to receive the bit of protoplasm from the end of the pollen tube, it consists of a large cell, called the *embryo-sac*, amidst a mass of cells. This mass is covered, except at the top, by two membranes called *micropyle.* The embryo-sac at this stage possesses two groups of nuclei one at the top of the sac close to the micropyle, one nucleus is slightly larger than the others, and this is called the *egg-cell.* The other two cells seem to be helpful in directing the contents of the pollen-tube to the embryo-sac. The three cells at the other or far end of the embryo-sac do not seem to take any part in the formation of seed. Within the embryo-sac, midway between the two groups of nuclei, lies a single large nucleus, which is called the nucleus of the embryo-sac, or the *secondary nucleus.*

The pollen-grain, deposited on the stigma, emits a tube and this comes to the top of the embryo-sac finding its way through the style and the micropyle. At the end of the pollen-tube we find two nuclei, and both of them get into the interior of the embryo-sac and one fuses with the embryo-cell and the other with secondary nucleus. This fusion of the nuclei of the pollen-grain with those of the ovule is really the process of fertilization. It is only after this fusion that the egg-cell is capable of division and development into the young embryo plant that we find in seed. The secondary nucleus divides and gives rise to the endosperm, after the fusion of the nucleii. Now it is obvious, that, for the production of offsprings, the fusion of the male and female cells is essential even in the case of plants, as it is in the case of animals. Further the offsprings are likely to be better in quality when the sex cells uniting together are from different plants.

The pollen-grains of a good many of the flowering plants have no power of spontaneous movement. So they have to depend upon some external agency for the pollination of their flowers. Even in the case of plants with bisexual flowers, extraneous aid is necessary for pollination. The transfer of pollen to the stigma from the anthers of the same flower, as well as from the anthers of a different flower on the same plant, is called *self-pollination.* Pollination of the stigma of a flower with the pollen from flowers of a different plant is called *cross-pollination.*

As a matter of fact, self-pollination occurs in several plants, especially annuals. In some cases flowers are self-pollinated regularly, and such flowers are usually small and inconspicuous without smell or honey. As examples for this, we may mention the Chenopodiums. We have also certain plants wherein self-pollination is made impossible in the earlier stages, but later, just before fading, this become possible. For instance, in the flowers of *Hibiscus micranthus* and *Abutilon indicum* the style branches project a little above the anthers just when the flowers open, and so the pollen cannot reach the stigma. It is evident that pollen from a different flower has to be brought and deposited on the stigmas by some agency, either wind or insects. If the stigmas fail to receive foreign pollen, they bend down so as to come near the anthers in the same flower, to be at least self-pollinated. In both these flowers, the petals close and press on the style branches and thus assist them in coming near the anthers. In some flowers, as in *Evolvulus alsinoides*, the style lies bent away from the anthers just, at first, but later it changes its position, so as to bring the stigma nearer the anthers to make self-pollination possible.

The flowers of *Mirabilis jalapa* are also adapted for self-pollination, when cross-pollination fails to occur. These flowers open towards the evening the emit a very strong scent. Just then, both the stamens and the stigma are far exerted, the stigma alone projecting above the stamens. Usually two species of moths visit these flowers and if they are not visited by any insect, the stigma manages to get self-pollinated either by the elongation of the stamens or by the slight bending of the style, so as to reach the anthers.

The flowers of some Compositae are specially interesting in that they have special adaptations in their floral mechanism to secure self-pollination, should cross-pollination fail to take place. For example, in the plant *Tridax procumbens*, a weed found everywhere, the heads have bisexual flowers in the central portion of the disc and the ray flowers are all female. The stamens are epipetalous and the anthers are adherent, so as to form a tube. The style is bifid and the branches are pressed together, in such a manner that the receptive surfaces are in contact and so not exposed. At first, when the flowers are in contact and so not exposed. At first, when the flowers are unopened, the style is short, but, as soon as the flowers begin to open, the anthers dehisce and the style elongates. So the pollen, lying in the anther tube, gets pushed out and remains at the top of the anther tube. The tip of the style may also carry a small amount of pollen, besides what may adhere to the hairs of the style branches on the outer surface. The style grows and after projecting a little above the anther tube, the branches separate and diverge, exposing the receptive surface. The stigmas thus exposed remain receptive for sometime so that cross-pollination may occur. If the stigmas fail to receive pollen by insect visits, the style branches diverge still more and even curl round, so that the receptive surface may come in contact with the pollen adhereing to the lower surface of the style branches, or with the pollen lying at the top of the anther tube. Self-pollination is thus ensured, if cross-pollination has not already taken place.

Instead of this makeshift arrangement to secure self-pollination, when cross-pollination is a matter of some difficulty, some plants produce two distinct sets of flowers, one adapted for cross-pollination and the other for self-pollination. In *Commelina benghalensis* we have a plant of this sort. The beautiful blue flowers are regularly cross-fertilized and the underground or law-lying inconspicuous flowers that remain as buds without opening are self-fertilized. These inconspicuous flowers are called *cleistogamous* flowers.

Charles Darwin, the great English naturalist, has proved by a series of experiments with different plants that cross-pollination is more advantageous to a plant than self-pollination. If we examine flowers of a number of wild plants growing in a place, we find in them numerous contrivances favouring cross-pollination to the exclusion of self-pollination. Recent workers in the field of plant-breeding have also established beyond doubt, that cross-pollination is not only more advantageous to the plant than self-pollination, but it also confers on plants certain racial advantages.

A most perfect arrangement to ensure cross-pollination is to have the essential organs on separate flowers. We find a host of plants in which the flowers are unisexual. In some plants, as in Cucurbita and Ricinus, we find both male and female flowers on the same plant (monoecious), in others, as in *Cephalandra indica* and in some palms, the male and female flowers are found on different plants and not on the same individual (dioecious).

Bisexual flowers become adapted for cross-pollination to the exclusion of self-pollination, by having the essential organs active at different times. The stigmas may nature and become receptive prior to the anthers. In Cumbu spikes (*Pennisetum typhoideum*), the stigmas protrude from the spikelets and become receptive, while the anthers are still within the glumes. Such flowers are called *protogynous.* The female sexual organs in the protogynous flowers are ready for fertilization long before the anthers are ripe in the same flower, and, therefore the flowers have to be pollinated only by the pollen of older flowers. There are also plants wherein the stamens shed their pollen prior to the receptivity of the stigmas. The flowers are then termed *protandrous.* Protandry seems to be more common than protogyny. Flowers of several Compositae, Malvaceae and *Andropogon*, *Sorghum* are protandrous. Pollination in this case is effected by the pollen from younger flowers.

In very many plants self-pollen is sterile, as in the case of many leguminous plants. The pollen in the flowers of certain orchids is said to act as a poison, if it falls on the stigma of the same flower. There are also instances in which the pollen from the same flower fails to fertilize, if foreign pollen falls on the stigma, soon after self-pollination. This means that foreign pollen is prepotent over self-pollen.

Another very effective arrangement promoting cross-pollination exists in the flowers of several species of Jasminum. The corolla in all cases is tubular and the stamens are epipetalous. In some flowers the style is short reaching only hair the length of the tube, whereas the anthers are at or near the throat of the corolla. On the same plant we also find flowers in which the styles are longer and coming up to the throat of the corolla, while the stamens remain within the tube about midway.

In the case of land plants it is obvious that for cross-pollination some extraneous aid is necessary, because of the fixity in position of these plants and of the absence of spontaneous movement on the part of the pollen-grains. The agents usually active in this work are wind and insects and, in rare cases, water. Plants depending upon insects for fertilization are very many and those pollinated by wind are not inconsiderable.

When plants depend upon wind for pollination, the stigmas have to wait for the pollen that may be wafted by the wind, and this is purely a matter of chance. Therefore, we should expect to find certain conditions specially favourable to this process in these plants. In the first place, large quantities of pollen should be available to ensure pollination. Winds may carry large quantities of pollen, but the pollen likely to fall on the stigmas can only be proportionately very inconsiderable. And, therefore, the flowers that are to be wind-pollinated should stand out far above the foliage leaves, so that the pollen may not be hindered from reaching the flowers. This is exactly what we find in the case of wind fertilized plants. For instance in grasses which are wind-pollinated the inflorescence rises far above the level of the foliage; in the case of trees dependent on wind for pollination, flowers appear, in most cases, at a time when the leaves have all fallen. As an example we may mention the tree, *Odina Wodier.*

The abundance of pollen is secured either by increase in the number of male flowers. Sometimes the anthers become larger and produce plenty of pollen. To give a vivid idea of the abundance of pollen produced by this class of plants, we may mention the male spikes of Pandanus. A single plant is capable of producing a very large quantity of pollen. It rate sized *Zea mays* plant produces about 50,000,000 pollen grains. The pollen produced by these plants should also be adapted for being carried by the wind. So these grains are smooth, light and dry. The anthers are in most cases versatile, an arrangement best suited for the shaking out of the pollen very readily.

The stigmas also have certain adaptations so as to enable them to catch the stray pollen floating in the air. They are branched and plumose, thus getting a large amount of surface. All grasses and sedges, some Amaranths, Ricinus, Pandanus and Odina are wind-pollinated. From this we see that wind-pollination involves the expenditure of a large amount of energy. As the pollen grains are composed almost wholly of protoplasm, a material most difficult to manufacture, this enormous production of pollen-grains cannot but be a drain on the resources of the plant. But this is necessary for ensuring pollination.

We have now to consider plants that are pollinated by insects. For successful cross-pollination flowers should be visited regularly and systematically by insects. Casual and erratic visits, or indiscriminateness in the chcice of flowers for visits are not likely to be beneficial. Insects should have some inducement to visit the flowers. Flowers, as we know., produce pollen and honey, and both these substances are sufficient inducements to attract the insects and also ensure their visits. By the mere secretion of honey in the flowers the insects cannot be allured. To ensure their visits, it is absolutely necessary that the place where honey is secreted should be made known to them. The colours of flowers are meant to show them the place where honey may be obtained.

The infinite variety in colour and form of the flowers, as well as the different scents, is intended to attract the insect. In many plants the petals or the corollas are large and coloured so as to be very conspicuous. If the flowers are small, they are rendered conspicuous by being massed together. Sometimes instead of the petals, bracts play the same part, as in Bougainvillaea. Stamens also partake in this work, in Neptunia and Dichrostachys. There are also instances of flowers wherein the calyx becomes highly coloured, while the corolla is absent. The brightly coloured part in the flowers of Mirabilis and Boerhaavia is the calyx. Even when the petals are present the sepals become coloured like the petals, and were have such examples in Cassia and Caesalpinia.

The colour and the scent of the flowers are no doubt quite sufficient to attract the attention of the insects, but to ensure the regular visitation of insects this is not enough. Something more substantial should be offered to them to induce them to frequent the flowers very regularly. The pollen serves as food, or as material for building the hives for some insects such as the bees. Further, nectar or honey is also secreted in several flowers. Insects such as moths and butterflies live mostly on nectar. As both these substances, honey and pollen, serve as food for these insects, they come to the flowers.

Insects that visit the flowers are bees, butterflies, moths and flies. All these insects have sucking apparatus, which in the case of flies are sort and in moths and butter-flies they are very long and coiled like a spring. Bees take in honey and also gather pollen; moths, butterflies and flies drink only honey. As stated above, these insects visit the flowers, because they get food in the form of honey or pollen. Therefore, the secretion of nectar or honey is essential to ensure their visit. In the case of insect-pollinated flowers, pollen need not be abundant. So, in these flowers, stamens are generally fewer in number. Further, the grains have usually a rough surface which often also becomes sticky. Stigmas need not have a large amount of surface as in the case of wind-pollinated flowers. The special adaptations in these flowers are so perfect that pollination is a certainty.

The nectar or honey is secreted usually by a disc at the base of the ovary. There are also instances of petals and sepals having nectaries either as glands, or as special pouches, sacs or spurs. The position of the nectary and the arrangement of the other parts of the flower are such as to make it impossible for the insects to get at the honey, without at the same time effecting pollination. It must also be remembered that, is the visit of the insect is to be of use in pollination, it is absolutely necessary that the part of the body of the insect which comes in contact with the anthers should also come in contact with the stigma ; otherwise pollination will not take place.

In flowers in which the corolla is regular and shallow any insect can get at the honey and effect pollination. For instance, in the Tribulus flower there are honey glands at the base of the stamens and between the petals; the coralla is shallow and the stamens stand erect and so any insect, a bee, or butterfly or a fly can get at the honey. In trying to get at the honey the lower portion, or the sides of the insect must necessarily get dusted with the pollen, and as the stigma is also of the same height as the stamens it must also touch the same part of the body of the insect.

Flowers having long tubular corollas like those of Vinca and Jasminum, or papilionaceous corollas like those of Phseolus, Indigofera and Vigna are not meant to be visited by all kinds of insects. Only butterflies and bees are able to get at the honey and in their manoenvres in search of honey they cannot help pollinating the flowers. Other insects cannot get at the honey and so they do not visit these flowers. Colour of the petals through intended to show off well by contrast, is probably not capable of attracting insects as well as the scent of flowers and the smell of honey.

The Mirabilis flower affords an example in support of this statement. This flower is visited by one or two species of the hawk moths; and there are several varieties of Mirabilis and in all of them the flowers open towards the evening and emit a strong scent. Sometime after sunset the flowers that are white, light yellow or cream-coloured alone can be seen and the magenta coloured flowers cannot be seen. Yet the moth visits both the white and red flowers with equal ease. Insects cannot see the colour from a distance, their range of vision in this respect being limited to within a few feet. Again they are able to see some colours better than others. Bees are said to perceive blues better than yellow. Scarlet does not seem to attract them. Many of the insects visiting the flowers are usually busy with only one kind of flower at a time.

For instance the common Carpenter bee was seen while sucking honey from *Bauhinia tomentosa* flowers. For over quarter of an hour it continued to hover about the same flowers, although there were several other flowers close by, that are usually visited by these bees. On another occassion this bee was thus visiting only the flowers of *Dolichos Lablab*. Having considered the question of pollination in a general way, we shall now deal with a few flowers in a more detailed manner. The flowers of Papilionaceae are specially adapted for cross- fertilization. At the base of the ovary there is usually a disc, which secretes plenty of honey. As the stamens are united by their filaments the honey finds its way into the staminal tube. Of the five petals, the standard is the chief

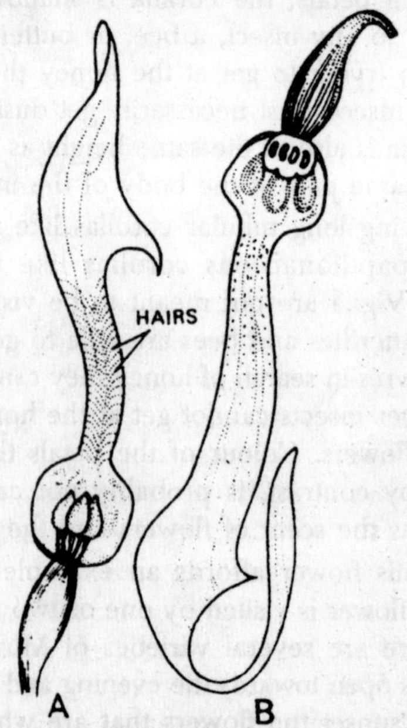

Fig. 2.8. Aristolochia showing pitfall mechanism.

attractive portion. It stands erect and a very conspicuous; in some cases besides being conspicuous it also bears some special marks which are supposed to serve as guides to the place where honey is secreted. The wing-petals invariably constitute a platform for the insects to alight and move about. The insect always sits on the wing-petals with its head directed towards the standard, whether there are special marks or not in it to direct the insect to the honey. Getting at the honey in these flowers is not at all an easy affair. It lies hidden at the base of the ovary and in the trough or cavity formed by the filaments. So it is only very intelligent insects that are likely to get at the honey.

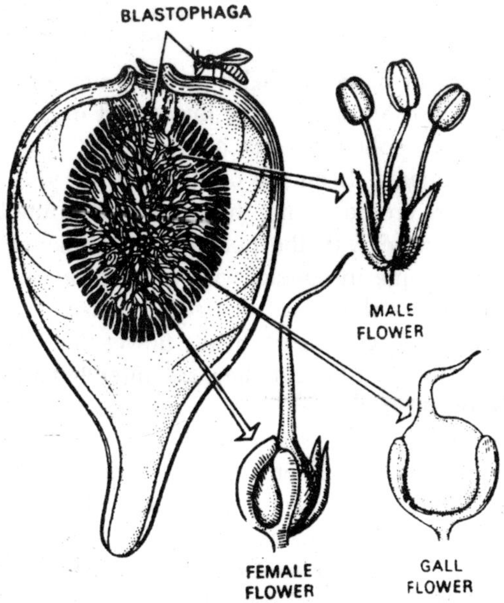

Fig. 2.9. L.S. of Ficus showing orific and Blastophaga.

When an insect settles on the flower and moves about, either the stamens or the stigma come out and touch the body of the insect. In some cases, as in *Crotalaria verrucosa* and *Tephrostia maxima*, when insects alight on the wing-petals and move about, pollen-dust is rejected from the pointed tips of the keel-petals first and then the stigmas come out when the insects visits the flower later.

In the flowers of *Vigna Catiang* the filaments of nine stamens are united, one stamen alone being free. At the base of the ovary and around it there is a disc secreting honey. The petals are all connected at the base by means of outgrowths, dents and foldings in such a manner that the standards stands erect, the wing-petals spread themselves out horizontally so as to serve as a platform and the keel-petal remains below. The margins of the keel-petals are united, both above and below, except at the free end where there is a small opening just enough for the emergence of the stigma. When the wing-petals are pressed down, the keel gets depressed and the stigma pops out, and when the pressure is removed, it resumes its original position. In this flower the anthers shed their pollen, before the stigma becomes receipt. Although the stigma is close to the anthers, the pollen does not reach it, because of the hairs found on the style a little below it. Even if a few grains fall on it they will not germinate, as the stigma is not receptive.

When a bee or butterfly alights on the wing-petals the stigma comes out and brushes against the lower part of the body of the insect. If there is pollen already on the body of the insect the stigma gets pollinated. If on the other hand the visit is at a time when the anthers are dehiscing, the lower part of the body of the insect will be dusted with the pollen, and the same part of the body is sure to come in contact with the stigma when it goes to another flower. In the plant *Indigofera enneaphylla* the floral mechanism is of a peculiar sort. In the flower the keel-petals have short spurs at their sides and they are intended to support the wing petals and keep them in a horizontal position. A very slight pressure on the wing-petals is enough to depress the keel-petals and separate their upper margins. As soon as the keels get depressed, the stamens and the style seem to come up with a jerk and remain outside. The essential organs cannot regain their original position, because the petals fall off as soon as the flower has received a single visit. So in this case one single visit of the insect is enough for cross pollination. Instances of the same kind of explosive arrangement are afforded by the flowers of *Abrus precatorius* and *Alysicarpus rugosus.*

The flowers of Phaseolus are very interesting in their structure and behaviour. Like other papilionaceous flowers these also depend on insects for pollination; at any rate the arrangement of the petals and the position of the essential organs are such that self-pollination cannot take place. The keel-petals are in this flower prolonged into a beak and the beak is in the form of a spiral. The end of the spiral in the flower of *Phaseolus trilobus* and *P. mungo* is towards th right, looked at from the front. The right keel-petal possess a spur which supports the right wing-petal and helps it is retaining a horizontal position, so that it may serve as a platform for insects visiting this flower. There is a distinct passage between the right wing petal and the end of the keel-spiral, leading to the base of the standard where the honey is found. The left wing-petal lies higher than the right over the keel on the left side. An insect can get the honey only from the right side. On the left side the wing-petal, the keel and the standard are all close together and there is no opening. An insect coming to this flower for the sake of honey will, of course, alight on the lower wing-petal on the right hand side. As soon as it alights on the wing and begins to search for honey there will be some pressure exerted on the wing-petal. The keel will also be depressed on account of the spur of the keel on the right side. In as much as the insect is on the platform, the stigma touches the back part of its body. If the stigma is receptive and if the back of the insect's body is already dusted with pollen, pollination takes place. If on the other hand the insect visits a flower for the first time to start with, it will get dusted with the pollen, and the next flower visited will have its stigma pollinated.

In *Leucas aspera* and *L. linifolia* the corolla is bilabiate and the stamens are epipetalous and didynamous. The stamens and the style are enclosed within the upper lip of the corolla. Honey is secreted at the base of the ovary and when the insect alights on the lower lip of the corolla to such the honey, the stamens and the stigma are released from the upper hood of the corolla. As both the anthers and the stigma are turned downwards they will touch the back of the insect; if the insect is already dusted with the pollen then the stigma will be pollinated, and in case the visit is for the first time the back of the insect gets dusted with pollen.

The flowers of acanthaceae have also special devices for pollination. In many plants of this family the stamens lie within the corolla in such a way that they cannot help coming in contact with the back of the insect when it sucks honey. Even small inconspicuous flowers are regularly visited by butterflies. On one occasion in the month of October while engaged in collecting plants, the tiny flowers of *Justicia procumbens* were constantly visited by the small pretty violet butterfly, Zizera lysimon. At first the flowers of *Justicia procumbens* were not noticed; this small butterfly was seen hovering about and settling on this plant. On close inspection it was found to suck honey from the tiny flowers of the Justicia. These plants were not at all numerous and there were stray plants scattered all over the place. The violet butterfly continued to visit these flowers persistently for over twenty minutes. The insect proved a most satisfactory guide to lead to the places where this species of Justicia was growing. Similarly another butterfly, a white one with a red border on the front wings, was seen visiting the flowers of *Spermacoce hispida* with great assiduity. The flowers of *Tephrosia purpurea* were visited by a butterfly with golden yellow wings. The flowers of Tridax were watched to see what insects visit them for honey and pollen. Seven different species of insects were found visiting these flowers and they were Zizera lysimon, Colotis amata, Catopsila pyranthe, Terias hecabe, Parnara mathias and Ceratina viridissima.

There are flowers that are exclusively pollinated by the visits of flies. The flowers of Aristolochia and Aroideae are of this kind. A very foetid odour is usually emitted by some of the Aristolochias and in one species the smell is so strong as to attract flies from very long distances (fifty to hundred yards or more). In this flower the anthers are situated on the style below the stigma and they dehisce after the withering of the stigma. Inside the tubular perianth, when it opens there are found a number of hairs all directed inwards. The flies get into the perianth tube easily. But they cannot get out of it with the same ease, on account of the hairs. This flower is protogynous and therefore it has to be pollinated by pollen from some other flower older than itself. When the flies get inside the flower they pollinate the stigma if they have already visited other flowers; if on the other hand the visit of the insect is its first visit, it gets dusted with the pollen. When the flies get into the flower, the anthers are usually immature and the flies cannot get out of the flower, until the anthers dehisce, because the hairs dry only then. Until that time the flies will be wandering all over inside, in search of a way to escape; and in their wanderings they get dusted with pollen. By the time the pollen is liberated the hairs disappear and stigmas fade, and the way becomes clear for the flies to escape.

The inflorescence of Aroids are particularly interesting on account of the arrangements existing in it for pollination. The spadix bears the unisexual flowers at its lower portion only, those at the very base being female and those above male. Above or between these are a number of reduced flowers in the form of hairs directed downwards. The whole inflorescence is covered by a large spathe, the lower part of which forms a tube constricted a little above the region of anthers, or the hairs, whilst the upper portion is expanded exposing the upper free end of the spadix. Small flies attracted by the bad smell emitted by the flowers and also by the spathe, crawl down into the tubular part of the spathe. Once they are in, they cannot easily get out from the tubular part, because the neutral flowers are all directed downwards and the spathe is constructed just above these flowers, or the anthers. The flowers are protogynous as in the case of Aristolochia, and so the insect pollinates the stigma if it bears pollen from another plant. If the flower happens to be the first one that is visited by the insect, the insect is forced to stay within the spathe until the way is open for it to escape.

The withering of the downward directed hairs and the dehiscence of the anthers take place together and therefore the flies escape with a coating of pollen. They cannot avoid this as the anthers are at the top of the inflorescence just at the level of the constriction or a little below it. In the flowers of orchids we have a most perfect mechanism for securing cross-pollination although it is very complicated. As examples we shall select two species of orchids commonly met with in the plains.

The orchid *Eulophia virens* grows amidst scrubs and its flowers are greenish with red streeks. There are three sepals and three petals; the sepals are all alike, but one of the petals, the one lying on the lower side of the flower when it is open, is modified, and it is called the lip. In this flower the lip is slightly saccate at its back and the free front part is broad and in the flower it stands out as a platform for the insects to sit on. The short shallow sac secretes honey and the flowers are visited by butterflies. The anther is single and it is on the top of a column rising from the top of the ovary. The pollen in this and all orchids is collected in masses called *Pollinia*, and in this flower there are two such masses in the anther. Just below the anther there is a disc-like gland to which the two pollen masses are attached by means of strap-shaped stalks. The anther in the open flower lies just above the opening leading into the honey-sac. When the butterfly sucks honey the pollen masses stick on to its head, and so the insect carries away with it the pollen masses. The next flower that is visited by this insect is pollinated by the pollen masses sticking on to this head. The stigma lies just below the anther on the front side of the column. When the insect rushes into the flower to such honey the pollinia cannot help touching the stigma because the pollinia assume a direction which will coincide more or less, with the position of the stigma.

In Habenaria we have another very common orchid of the plains. The flowers in this orchid are all pure white and the raceme springs from the middle of two or three elliptic leaves laying flat on the ground. They are very conspicuous by reason of the recemes. Both in the morning just before sunrise and in the evenings the flowers are visited by moths. The lip is prominent below and it is prolonged at the back into a very distinct spur varying in length from quarter of an inch to one inch. The two lateral petals and the sepal lying above are united so as to form a hood.

There is a narrow passage leading into the interior of the flower and the anther is situated on a column just above the passage leading into the spur. The pollinia have long stalks and the basal portions of the stalk protrude a little. When the moth gets in, the two pollinia stick on to the two eyes of the insect and when it flies to another flower the pollinia are rubbed against the stigma lying below the anthers and the masses adhere to the stigma as its surface is very sticky. In *Calotropis gigantea* (as also in all Asclepiadeae), the pollen grains are held together in masses and

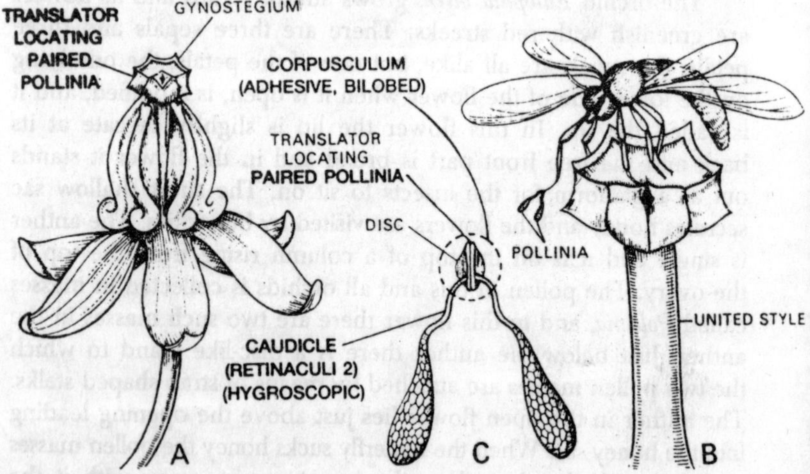

Fig. 2.10. Pollination in Calotropis. A–A flower of Calotropis, B–Bee removing part of pollinia, C–A pair of pollinia of Calotropis.

there are ten pollinia. When an insect settles down on the top of the staminal column in the centre of the flower and moves about round the stigma, the pollinia adhere to the legs of the insects. When it flies to another flower the pollinium gets on to the stigma to which it will stick if it is receptive. There are many plants whose flowers are usually-pollinated by moths. As moths move about only during the night or in the dusk, these flowers have white corollas, open in the dusk and also emit a strong scent at this time.

The flowers need not be irregular as a platform is not needed, because moths do not alight and they only hover in front of the flower. We may mention as examples Ipomoea, Tobacco, Mirabilis and Jasminum. In these flowers the anthers are generally loose and are well adapted to shed pollen on to the body of a hovering moth. The hawk moths, *Herse Convolvuli* and *Cephonodes picus* visit the flowers of many of the large flowered ipomoeas and the flowers of *Mirabilis jalapa.* The former is a very constant visitor, but the other one comes very rarely and it is extremely active and so rapid in its movements that it is impossible even to see it. We have also plants that depend on water for pollination and they are all aquatic plants.

The common water plant, Vallisneria may be selected as a type of plants fertilized by the agency of water. The flowers of this plant are unisexual and the plants are dioecious. Both the staminate and the pistillate flowers develop under water, and so, unless there is some special adaptation, fertilization cannot take place. The male flowers are borne on short stalks almost at the base of the plant. At the time of flowering, or when the flowering season approaches, these get separated from their stalks and rise to the surface of water to float there. Simultaneously, the female flowers lying under water with spirally coiled stalks come up to the surface, in consequence of the uncoiling and growth of the stalk.

The male flowers are most numerous and they float amongst the female flowers. The anthers open and the pollen is shed on the stigmas. In the case of flowers that are hermaphrodite, the flowers are raised above the water level, as in Nymphaea, so that they might be pollinated either by insects or by wind. It is only when flowers are unisexual and also submerged, pollination is effected by water currents. In all plants that are pollinated by currents of water the pollen grains are smooth, and pollination also is not a matter of certainty. So in these cases there is an abundance of pollen. During hot weather when water is low in ponds and tanks we generally see masses of small white flowers floating freely on the surface. Sometimes the acquatics which ordinarily raise their flowers above the water, do not do so when rain is heavy. Then the flower buds do not open and are then cleistogamic.

Protection of Pollen against Rain

Microspores are much less resistant to extremes of temperature and to drying when they have been moistened and in consequence have begun to germinate. Pollen may be protected from rain in various ways. In some flowers, especially those whose pollen is exposed to rain when the flower opens, the microspores are not readily wetted, having a covering of wax or of spines, etc. Many flowers protect the pollen by their horizontal or drooping position *e.g.*, Ericaceae, Liliaceae. In some cases the flower closes up at night or in bed weather, *e.g.*, *Oxalis*, tulip, crocus, some Ranunculaceae; and the same kind of closing is effected in the capitula of many Compositae by the movement of the flowers and bracts. In iris the large petaloid stigmas cover the stamens, and in many flowers the stamens are protected by a hood formed by the sepals or petals, or by both.

Germination of Microspore

Process leading up to and ending in Fertilization. At first the microspore is unicelllular but later, even before it leaves the anther, its nucleus divides into two. The larger one of those in the vegetative nucleus; the smaller one the generative. Both are free in the cytoplasm of the microspore. Either before or after pollination the generative nucleus divides into two male gametes. Germination and further development take place on the stigma, which secretes a sugary nutritive fluid. From one of the pores in the exine a slender pollen-tube protrudes, and grows down through the tissue of stigma and style and finally enters the ovary. After entering the ovary the pollen-tube grows towards and ovule, which it enters usually by the micropyle. It penetrates the apex of the nucellus and comes in contact with the megaspore near to the oosphere and synergidae.

During its growth the vegetative nucleus is found near the tip of the pollen-tube. The male gametes also pass down to the apex of the pollen-tube, but the vegetative nucleus is by this time disorganised. *One male gamete only* is concerned ir. the actual process of fertilization. It passes from the pollen-tube into the megaspore and fuses with the oosphere, hence their names synergidae or "help-cells" (Gr,*ouv*, with *epyov*, work). The tip of the pollen-tube swells and bursts, thus setting the male gametes free in the embryo-sac. The fusion of the male gamete, which appears to be entirely nuclear, with the nucleus of the oosphere constitute fertilization in the strict sense. The fertilized oosphere secretes a cellulose wall and is then called the oospore. Its nucleus is diploid.

The second male gamete fuses with the secondary nucleus of the megaspore. The resulting nucleus is called the *endosperm nucleus*. The significance of this process, which resembles fertilization, and which, together with the actual fertilization of the oosphere, constitutes what has been called "double fertilization," is considered in § 10. In a few Dicotyledons, *e.g.*, *Corylus* and *Betula*, the pollen-tube does not enter the ovule by the micropyle, but by piercing the chalaza. This is known as chalazogamic fertilization as distinguished from the usual porogamic method.

3

Fertilization

Fertilization involves the fusion of a male gamete with a female gamete. In angiosperms the female gametophyte is seated deep in the ovarian cavity, quite away from the stigma. The pollen (male gametophyte) are, normally, held at the Stigma, and there is no device for them to reach the egg inside the female gametophyte. To effect fertilization in this group of plants the pollen grains germinate on the stigma by putting forth tubes (pollen tubes) which grow through the style and find their way into the ovules, where they discharge the sperms in the vicinity of the egg. One of the sperms fuses with the egg (forming zygote) while the other fuses with the polars or the secondary nucleus (forming primary endosperm nucleus). The distance a pollen tube has to travel in order to reach the egg depends on the length of the style which is quite variable in different species. For example, in sugar-beet it has to grow only a few millimetres whereas in corn it grows as much as 450 mm.

Structure and Development of a Stamen

A very young stamen in a flowers bud is small, four angled, upgrowth from the receptacle. This upgrowth will four the anther, the filament being delayed in development. In each of the four angles a pollen sac containing pollen grains will be elaborated as development proceeds. These developments as seen in transverse sections of anthers of various ages, but it must be realised that logitudinal series of cells are involved, running the whole length of the anther. The left-hand figures represent early stages. The anther consists at first of undifferentiated parenchyma but soon cells in

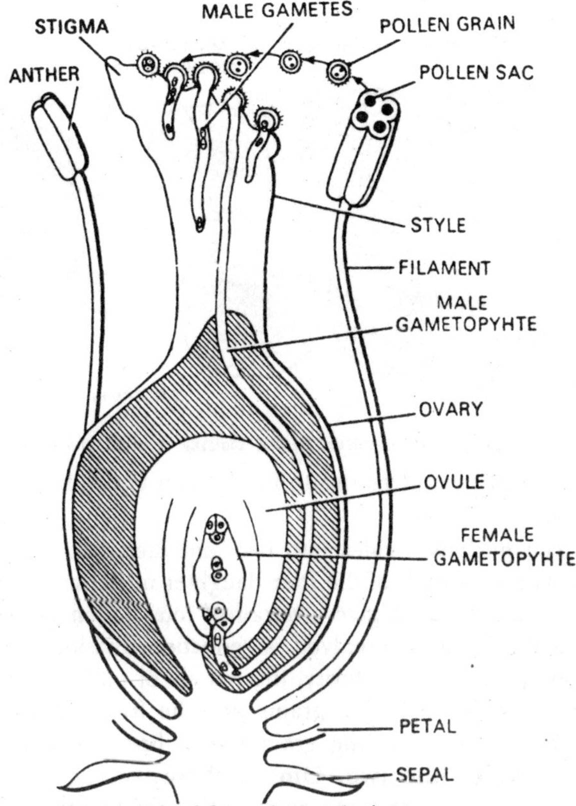

Fig. 3.1. L.S. of flower showing fertilization.

the four corners, lying just beneath the epidermis, become active. Divisions take place giving a group of inner cells (shaded) surrounded by flattened, actively dividing cells. The inner cells ultimately give rise to the pollen grains while the outer cells form the walls of the pollen sacs.

Division continue in both the inner and outer cell groups and the condition seen in the top right hand figure is attained, where each corner of the anther is occupied by a young pollen sac. Each pollen sac has a wall several layers thick, the innermost being a layer of cells known as the tapetum, which later provides food for the developing pollen grains. The cells in the centre of the sac are pollen mother cells with dense cytoplasm and large nuclei. These, at first joined closely together, soon round themselves off and float freely in a nutritive liquid derived from the tapetum. Each pollen

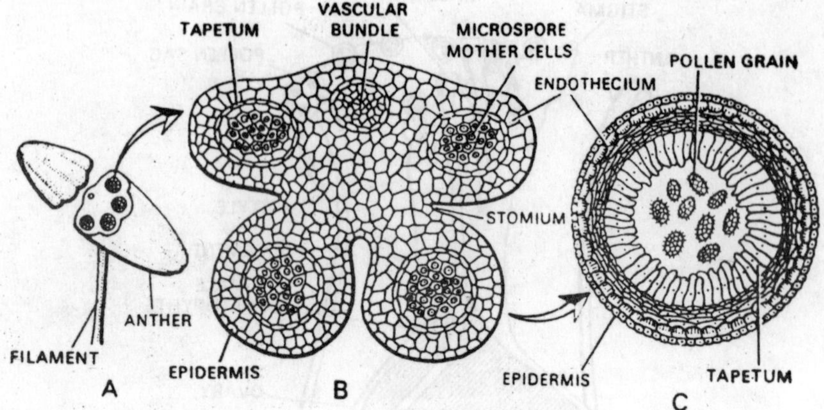

Fig. 3.2. T.S. of a mature anther, showing stomium and pollen grains.

mother cell now gives rise to four pollen grains by two successive divisions.

The nucleus divides into two and then into four. Cytoplasm collects round each of the four daughter nuclei and partition walls are formed so as to give a tetrad of four pollen grains. The type of nuclear division involved is reduction division or meiosis. As the anther reaches maturity, the tetrads of pollen grains break up and the separate pollen grains now lie freely in the cylindrical pollen sacs. Each young grain has at first one nucleus but this soon divides into two. One of these, surrounded by denser cytoplasm, forms the generative cell (which ultimately gives the male gametes) and the other is the tube nucleus. The wall of the ripe pollen grain is a double one. There is dedicated inner wall of cellulose (the intine) and a cuticularised outer one (the exine) of variable thickness. Thin places, or pore, are present in the exine. The grains vary in size from 3-300 µ. Their shape may be ellipsoidal, spherical crystal-like with a varying number of facts, or many other shapes. Further variation given by characteristic sculpturing of the outer wall which may have spines, papillae, a network of ridges and so on.

Many plants can, indeed, be identified merely on the basis of their characteristic pollen. It has proved possible to learn something about the changes which have occurred in our vegetation since the Ice Age by the identification of pollen grains preserved at various depths in peat deposits. It may also be mentioned that the earliest known trace of Angiosperms is in the form of pollen-grains found

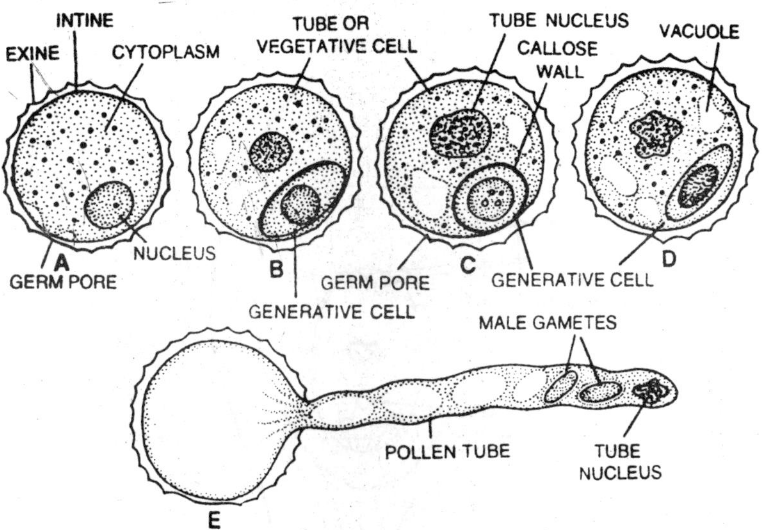

Fig. 3.3. Germination of pollen grain and formation of male gametophyte in an angiosperm.

in coal of Jurassic age in Scotland. As a result of the various developments described above the anther now has the structure. As the left hand figure indicates, the pollen sacs are long cylindrical structures. There are two of them in each lobe, the lobes themselves being joined by a connective containing a vascular bundle.

The wall of each sac has now a highly differentiated structure. The cells lying just below the epidermis have enlarged and their walls have been modified by the deposition of lignin. This characteristic fibrous thickening is laid down on all the walls except the outer ones. In some anthers several layers of the wall may show this type of thickening. The thin outer walls of the cells of the fibrous layer shrink on drying and so cause a longitudinal rupture of each anther lobe along a line of specialised cells called the stomium. This rupture is accompanied by a breaking down of the wall between each pair of sacs. The pollen grains are thus liberated and pollination can now take place.

Structure and Development of an Ovule

The ovule when it is ready for fertilization is more or less egg-shaped. It consists essentially of an ovoid mass of parenchymatous cells called the nucellus which is borne on a short stalk or funicle by means of which it is attached to the placenta in the ovary. The nucellus is invested by one, or more frequently

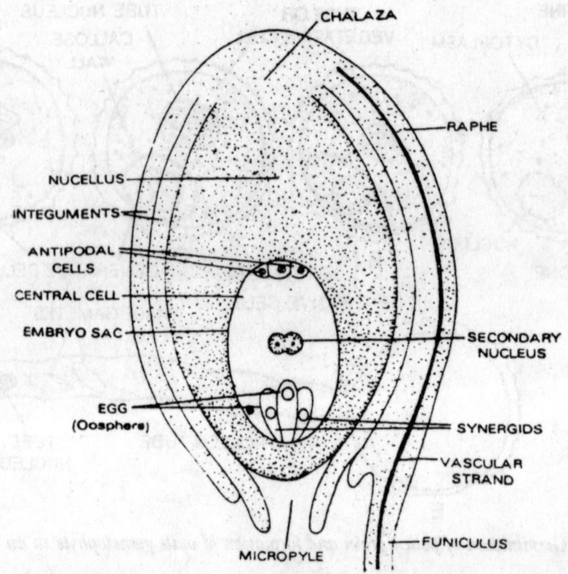

Fig. 3.4. Structure of a typical ovule (anatropous ovule) prior to fertilization.

two, integuments which are protective coats growing up from its base. These cover over the nucellus except at the extreme tip, where a narrow channel is left open as the micropyle leading down to the nucellus. Embedded in the nucellus there is a large ovoid sac bounded by a thin cell-wall. This is called the embryo-sac, since, after fertilization, it will contain the embryonic plant. In some plants, for example, rhubarb, the ovule is straight and the micropyle is at the end opposite to the stalk. Such ovules are said to be orthotropous. In other examples, for example, shepherd's purse, the nucellus and embryo-sac are curved so that the micropyle is placed near to the stalk. This is the campylotropous type of ovule.

In the majority of ovules, however, the nucellus is straight but the entire ovule is inverted by a sharp curvature at the top of the stalk so that the micropyle lies near to the base of the stalk and also near to the carpellary wall. This is the common anatropous type of ovule. The embryo-sac at the time of fertilization contains vacuolated cytoplasm in which a number of prominent nuclei are embedded. At the end furthest from the micropyle there are three large nuclei, each surrounded by granular cytoplasm and a gel membrane; these are called the antipodal cells and often appear to

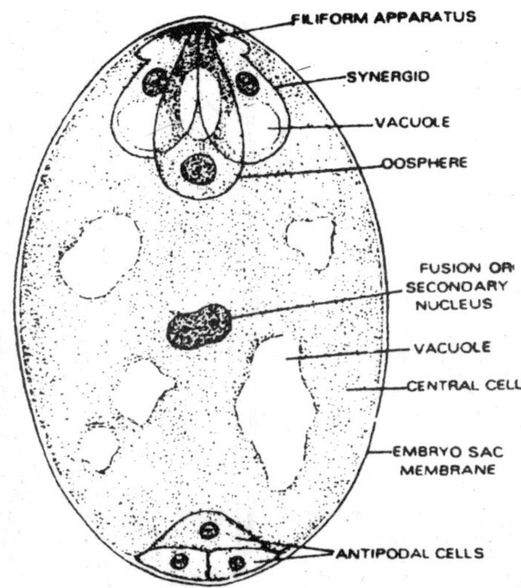

Fig. 3.5. Normal or Polygonum type of embryo sac.

play a part in the nutrition of the embryo-sac. At the micropylar end there are three other large nuclei and these too are each surrounded by cytoplasm and a membrane. This micropylar group constitutes the important egg apparatus. The central cell is the female gamete called the egg cell or ovum. The two other cells, which may correspond to non-functional ova, are called the synergids.

In the centre of the embryo-sac there is a conspicuous nucleus which is called the central fusion nucleus or the primary endosperm nucleus. When the ovule is in the condition just described it is ready for fertilization but before considering this process it is necessary to examine the way in which this structure has developed. An ovule arises on the placenta as a small upgrowth of parenchymatous tissue which is the young nucellus. Then integuments appear as ring-like upgrowths from its base and these gradually invest the growing nucellus until only the micropyle is left at its tip.

Meanwhile, if the ovule is of the anatropous type, the stalk has elongated and curved sharply at its top so as to bring the micropyle near the insertion of the stalk on the placenta. It is during these developments that the embryo-sac is formed. A single cell of the nucellus, situated just below its tip, becomes conspicuous by reason of its larger size and dense protoplasmic contents. This is the embryo-sac mother cell which then divides first into two

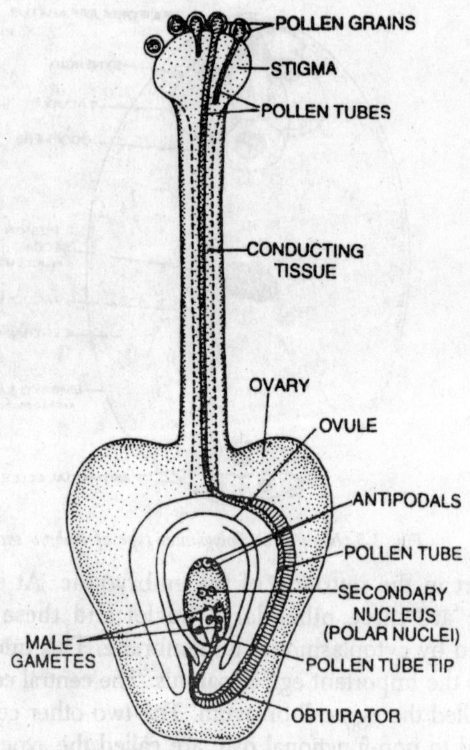

Fig. 3.6. Fertilization in an angiosperm through porogamy.

cells and then into a row of four cells. As in the formation of four pollen grains from a pollen mother cell, the nucleus of the embryo-sac mother cell divides by the process of meiosis to give the four potential embryo-sacs.

Usually, however, it is only the innermost of these which develops further as the functional embryo-sac. The other three collapse and are crushed as the surviving embryo-sac enlarges. When first formed, the embryo-sac has a single nucleus. This nucleus divides by mitosis and the daughter nuclei move to the two poles of the sac. Each of the two nuclei divides again twice so the four nuclei occupy the micropylar end and four the opposite (chalazal) end of the sac. One nucleus from each group of four now moves to a central fusion nucleus. The three at the micropyle end form the egg-apparatus already described and the other three form the antipodal cells. The embryo-sac thus reaches the condition which is found in practically all Angiosperms at the time of fertilization.

Processes of Fertilization

It is now necessary to trace the sequence of events following upon the deposition of pollen grains on the receptive stigma. The pollen grains germinate on the stigma where there is usually a liquid secretion, often containing sugar, but sometimes other substances are also present, as, for example, malic acid in the case of members of the Ericaceae. Germination results in the growing out of the thin inner wall of the pollen grain through one of the pores in the thick exine. This results in the production of a cylindrical pollen tube into which the tube nucleus and generative nucleus pass. The pollen tube, tending to grow towards chemical substances produced by the stigma, towards moisture and away from oxygen, then enters the stigma and begins to grow down the style. In some examples, for example, *Viola* species, there is an open canal in the centre of the style leading right down to the ovary; in others, for example, *Rhododendron* species, there is a canal

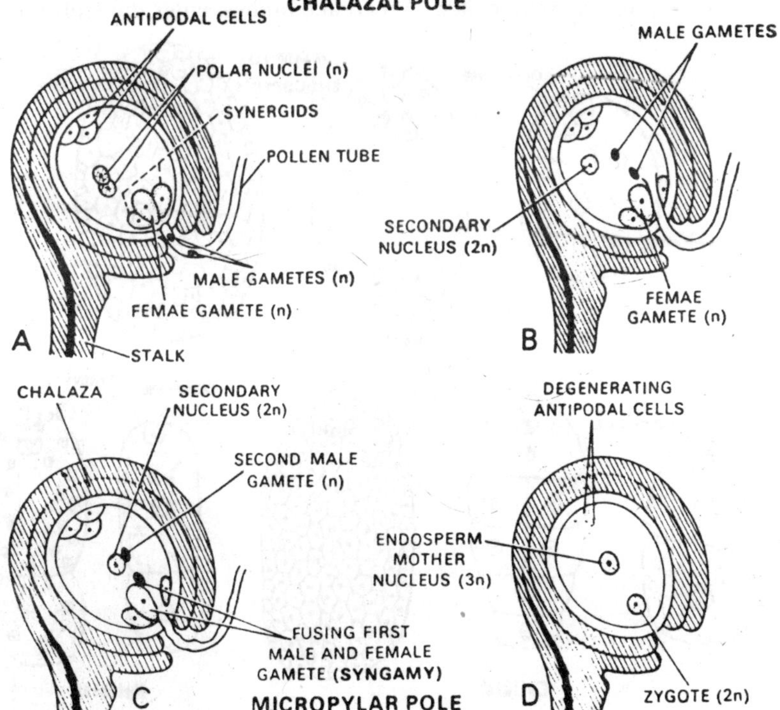

Fig. 3.7. L.S. of ovule showing different stages of fertilization.

filled with mucilage. The tube grows down the moist sides of the canal or through the mucilage, as the case may be, and so reaches the ovary. It obtains water and food for this continued growth from the tissues of the style. Having reached the ovary, the pollen tube either grows in the tissues of the ovary wall until the tip reaches the vicinity of an ovule when it enters the cavity of the ovary, or else it enters the top of the ovary cavity and grows down the moist inner surface in an ovule.

In either case the tip of the tube enters the micropyle of the ovule,. possibly directed by a fluid secreted by the synergids. It thus reaches the nucellus, and by penetrating the few overlying layers of this tissue, it comes into contact with the embryo-sac in the vicinity of the egg-apparatus. During the downward growth of the pollen tube the generative cell has divided to form the two male gametes. These occupy the tip of the tube, the tube nucleus having disintegrated. Fertilization now takes place.

The tip of the pollen tube becomes ruptured and the two male gametes, now spirally coiled, thread-like bodies, enter the embryo-

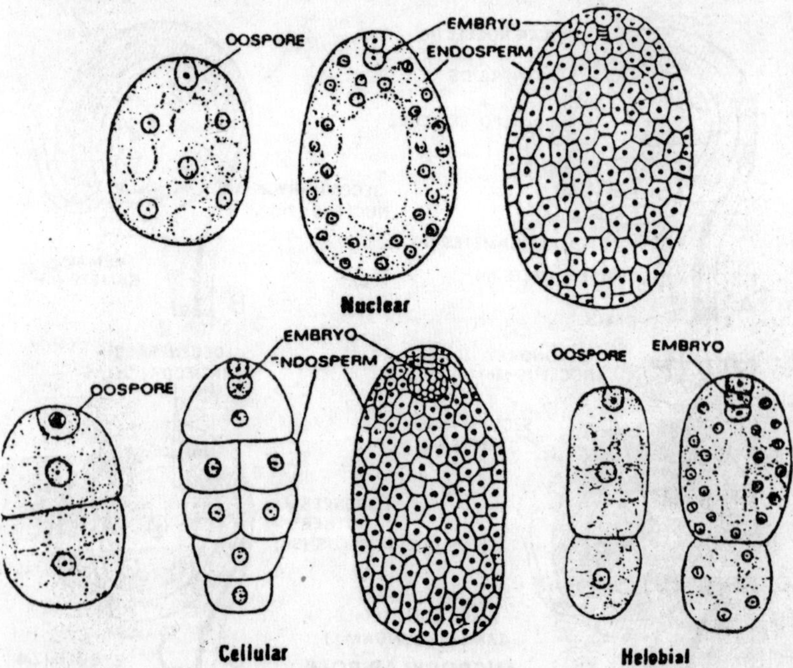

Fig. 3.8. Types of endosperm.

sac. One of the gametes enters the egg and fuses with its nucleus. The egg, so fertilized, is called the zygote and it is from this that the embryo plant develops. The second male gamete moves deeper into the embryo-sac and fuses with the central fusion nucleus. The result of this second fusion is the production of a nutritive tissue, known as the endosperm; this is used by the developing embryo. The time elapsing between pollination and the occurrence of these fusions varies considerably. In rye it is seven hours; in maize, twenty-four hours; in some trees it may be a year or more.

The elucidation of the sequence of the events just outlined occupied botanists for many years. Although it was known from the time of Camerarius onwards that a fertilization process did in fact take place, the details proved difficult to observe. In 1823, Amici, an Italian naturalist, observed the production of pollen tubes by pollen adhering to a stigma and seven years later he traced the course of the pollen tube down into the ovary and into the micropyle of an ovule. Later still (1846), the same observer showed that the ovule had an egg cell before fertilization and that this, excited by the pollen tube, gave rise to the embryo. Hofmeister confirmed these observations in 1849 but it was not until 1884 that Strasburger observed the fusion of one of the male gametes with the nucleus of the ovum.

It was still later (1898) before Nawaschin observed the fusion of the second male gamete with the central fusion nucleus in the centre of the embryo-sac. We have now seen that fertilization consists essentially of the fusion of a male gamete with a female gamete, the ovum, to form a zygote and ultimately an embryo. The second fusion is a subsidiary one and results only in the formation of nutritive tissue. In the account of fertilization given above the events have been described in relation to a single pollen grain and a single ovule.

Normally, however, many pollen grains will be deposited on a stigma and they will all produce pollen tubes. If the ovary contains only one ovule, naturally only one of these tubes penetrates the micropyle and liberates its male gametes into the embryo-sac. If, however, as in the majority of ovaries, numerous ovules are present, then the production of numerous pollen tubes is a necessity if all the ovules are to be fertilized. In a flower like that of *Rhododendron*, if a transverse section is cut through a style some time after pollination has occurred, the section will show a large number of pollen tube sections embedded in the mucilage of the stylar canal.

Meiosis

An essential feature of fertilization is the fusion of a male gamete with the nucleus of the female gamete. This results in the zygote or fertilized egg the nucleus of which contains the nuclear material of both the male and female gametes. If the male gamete has n chromosomes (a number varying with the species) and the female has n chromosomes the zygote will obviously possess $2n$ chromosomes. The chromosome complements in the vegetative cells of flowering plants consist of homologous pairs, one of each pair being maternal and the other paternal in origin. It is clear that these pairs came together in the act of fertilization which gave rise to the plant. As the zygote develops into the embryo, and then into the new plant, all the cell divisions which take place are mitotic and thus the $2n$ number of chromosomes is maintained. If, however, this type of division occurred throught out the life history, then the gametes would each have $2n$ chromosomes and, when fertilization occurred, the zygote would have $4n$ chromosomes. There is, however, a compensating process in the form of another type of nuclear division known as meiosis or reduction division, by which the number of chromosomes is halved.

Meiosis always occurs in the formation of the gametes or in some cell division prior to such gamete formation. The result is that the gametes themselves always have the haploid or n number of chromosomes and the zygote formed as a result of fertilization will have the diploid or $2n$ number. In flowering plants, meiosis occurs in the formation of the four pollen grains from each pollen mother cell and also in the formation of the four potential embryo-sacs from an embryo-sac mother cell. As the subsequent nuclear divisions which take place both in the pollen grain and in the functional embryo-sac are all mitotic, it follows that the make gametes and the female gamete will have the reduced or haploid number of chromosomes.

All plants which reproduce sexually show the occurence of meiosis at some point in their life-history between successive acts of fertilization but this point varies in the different big groups of plants. Meiosis consists of two very much modified mitotic divisions, normally giving rise to four cells, but it is essential that it should be considered as a single process with two stages. Cells which are about to undergo this type of division are usually filled with dense cytoplasm and have a large nucleus. The first indication that meiosis is in progress is the appearance of the chromosome threads just as in the prophase of mitosis, except that they are single and show the chromomeres more obviously along their length.

The stage where the strands are obvious as separate entities is called leptotene and this is followed by zygotene where the homologous chromosome strands come to lie together, so precisely paired that corresponding segments and even chromomeres are in juxtaposition. These paired chromosome structures are called bivalents. In each bivalent both of the chromosomes now split longitudinally (except in the region of the centromere) giving a four-stranded structure which is characteristic of the pachytene stage. Pachytene is a very short stage and is seldom seen in preparations showing meiosis; but it is nevertheless important.

In each bivalent the chromatids now contract and twist around one another but the members of each pair repulse the opposite members except at points which are called chiasmata, (sing., chiasma). These chiasmata explain cytologically the genetic problem of 'crossing over'. This stage is diplotene and it is followed by diakinesis, which is characterized simply by a greater shortening and thickening of bivalents. The latter usually arrange themselves around the periphery of the nucleus and this stage also the nuclear membrane and the nucleolus disappear.

Metaphase I follows and is similar to the metaphase of a mitosis in that a spindle has now been formed and that the bivalents arrange themselves on the equatorial plate in exactly the same way as do the chromosomes in mitosis. The bivalents are placed with their two unsplit centromeres above and below the equatorial plate. Anaphase I now follows and the two centromeres of each bivalent move to the opposite poles of the cell, each pulling its pair of chromatids with it. Telophase I can be a very transient stage, and can frequently be regarded as non-existent since the daughter nuclei pass straight into the prophase of the second division. Metaphase II follows, each centromere (of which there are only half as many as in the original cell) moves onto the equatorial plate, taking with it the two chromatids which are attached to it.

The centromeres then divide and at anaphase II the two chromatids separate. Four daughter cells are thus produced, each with half the number of chromosomes which were present in the parent cell. One feature requires fuller treatment. At diplotene there is actual breakage and interchange of parts of chromatids, the position of the breakages usually being seen as chiasmata. Chromosome which has one chiasma and the result of this chiasma as it effects the constittution of the segregating chromatids at anaphase I and anaphase II. The interchange of segments extends from the chiasma to the ends of the chromatids. This means that two of the chromatids are actually made up of both maternal and paternal material while the other two are either wholly maternal or wholly paternal in their make-up. More than one chiasma can be present in a bivalent and the interchange can take place between any two chromatids provided that one is of maternal and the other of paternal origin. Extremely complicated segregations can thus occur.

The Seed

After fertilization has occurred, changes take place in all parts of the ovule. The structure resulting from these further developments in the ovule is the characteristic reproductive body of the flowering plants, the seed. Changes in the embryo-sac itself may be considered first. The fertilized egg forms a wall around itself and proceeds to divide so as to form a short chain of cells which is known as the pro-embryo. The cell nearest to the micropyle grows to a large size, while the cell at the tip of pro-embryo begins to divide in various planes, these divisions following a regular sequence. This terminal cell in fact gives rise to most of the embryo plant, the rest of the pro-embryo forming the suspensor. The large basal cell of the suspensor serves as an organ of attachment to the wall of the embryo-sac; the filamentous region continues to elongate and pushes the embryo into the nutritive tissue, or endosperm, which is developing in the embryo-sac at the same time. Continued cell division in the embryo soon results in the appearance of two cotyledons (or of one lobe if the plant is a Monocotyledon).

The stem apex is formed between the two cotyledons (or laterally at the base of the single cotyledon in a Monocotyledon). The lower part of the embryo in both Dicotyledons and monocotyledons forms the bulk of the first root or radicle. The tip of the latter is, however, formed by divisions in the top cell of the suspensor. The tip of the root is thus attached to the suspensor and points towards the micropyle. Simultaneously with the development of the embryo, or even preceding this, endosperm is formed in the embryo-sac. This results from the active division of the central nucleus (which,

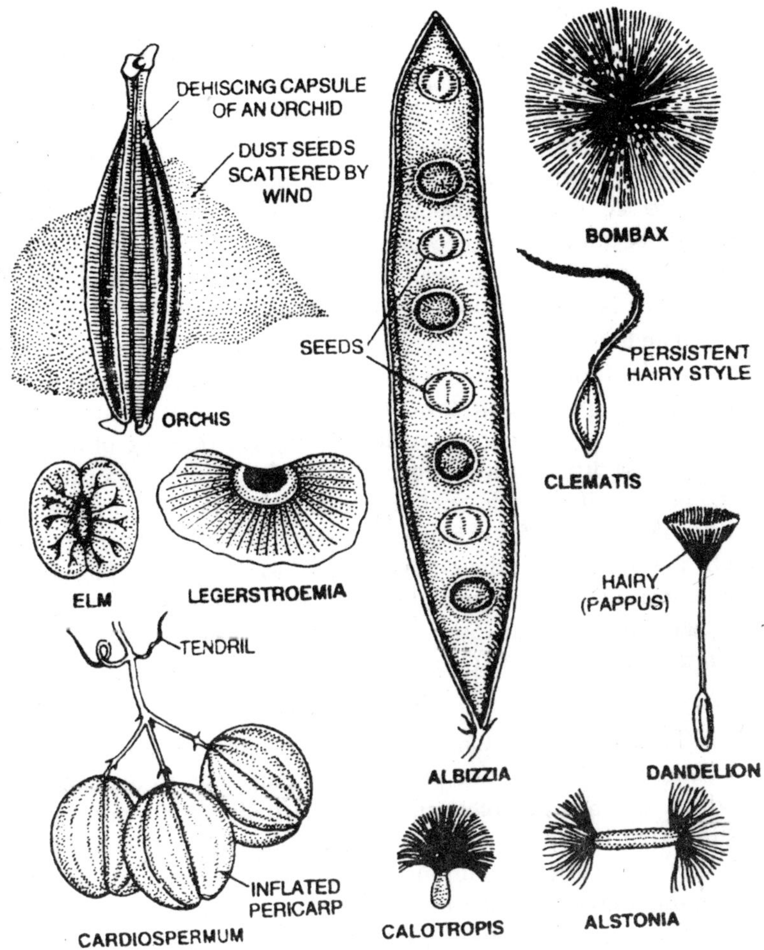

Fig. 3.9. Some examples of anemochorus fruits and seeds.

it will be remembered, is the product of the fusion of the second male gamete with the central fusion nucleus). Repeated divisions of this nucleus give rise to numerous daughter nuclei which are not, at first, separated by cell walls (a type of divisions known as free nuclear division). These nuclei take up a peripheral position in the embryo-sac. Continued unclear division takes place and then wall formation commences. This results in the formation of a cellular tissue which gradually encroaches on the cavity of the embryo-sac. This tissue, which is rich in food materials, is the endosperm. As the embryo and endosperm develop inside the embryo-sac the latter increases rapidly in size and, as a result of this enlargement, the tissues of the nucellus are crushed and practically obliterated. In a few examples, however, some of the nucellus persists as a nutritive tissue known as perisperm. The antipodal cells may soon disappear or, in other examples, they may persist and either enlarge or divide to form a tissue which helps in the nutrition of the enlarging embryo-sac. Changes also take place in the integuments. These usually become fibrous or woody and constitute the protective seed coat or testa. The total result of these post-fertilization chages in the ovule is the structure known as a seed.

The testa is, as just described, derived from the integuments of the ovule. Inside the testa is the embryo. In addition there may be endosperm present, or this tissue may be absent from the ripe seed. If the embryo grows rapidly during the ripening of the seed it will use up all the endosperm to obtain the requisite food material for its growth. Such a seed is said to be non-endospermic or exalbuminous (for example, the seeds of apple, bean, pea etc.). On the other hand, the embryo may cease to grow before all the endosperm has been used and then the ripe seed will be endospermic or albuminous as in castor oil (*Ricinus communis*), wheat (*Triticum vulgare*), and coconut (*Cocos nucifera*).

The Fruit

The effects of fertilization are not confined to the ovules but extend to the other floral parts and, in particular, to the carpels in which the ovules are borne. The carpellary or ovary wall is stimulated to further development, possibly by growth substances secreted by the developing ovules. It may become dry and membranous, or it may thicken considerably and become either woody or fleshy. This modified ovary wall, together with the contained seed or seeds, constitutes a fruit, the modified wall being the fruit coat or pericarp. It serves to emphasize the fact that the ovary wall becomes the pericarp (in this case, the pea pod), and that the ovules become the seeds. It may be noted that in a few cases, for example, banana (*Musa sapientum*), the fruit normally develops without fertilization having taken place. This phenomenon is known as parthenocarpy and naturally gives rise to a seedless fruit.

The development of seedless tomato fruits may be induced by spraying the flowers with a growth substance such as β-naphthoxyacetic acid. In this case the growth substance artificially

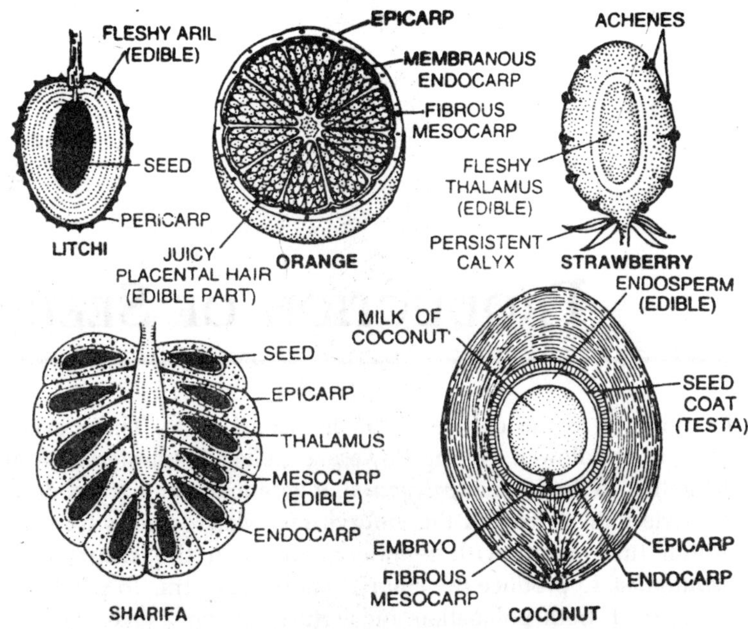

Fig. 3.10. Sections of some common fruits to show edible parts.

supplied seems to act in a similar way to the growth substance produced by ovules developing after normal fertilization. After fertilization the sepals, petals and stamens usually wither, although their dead remains may often still be seen, for example, at the top of an apple or gooseberry. In some examples, however, the sepals may be persistent (for example, pea, tomato). In the mulberry (*Morus* species) the perianth becomes pulpy and embeds the true fruits.

4

PRODUCTION OF SEEDS

The embryo is the end result of the sexual cycle. These events occur in the flower. During flowering, pollen is transferred from the anther to the stigma (*pollination*); it germinates and grows down the style until it reaches the embryo sac. Male gametes from the pollen tube unite with female gametes in the embryo sac (*fertilization*) to produce the embryo and endosperm. To produce a viable seed, both pollination and fertilization must take place. In some cases, however, the fruit may mature and contain only shriveled and empty seed coats with no embryo or with one that is thin and shrunken. Such "seedlessness" may result from several causes; (*a*) *parthenocarpy* (the development of the fruit without pollination or fertilization); (*b*) *embryo abortion* (the death of the embryo during its development); or (*c*) the inability of the embryo to accumulate the required food reserves. If embryo abortion occurs early, it is most likely that the fruit will soon drop or will not grow to its normal size.

FRUIT AND SEED DEVELOPMENT

Morphological Development

The relationship between the structure of the flower and the structure of the fruit and seed in angiosperms is as follows:

Ovary → fruit (sometimes composed of more than one ovary plus additional tissues)

Ovule → seed (sometimes coalesces with fruit)

integuments → testa (seed coats)

nucellus → perisperm (usually absent or reduced; sometimes storage tissue)

2 polar nuclei +
sperm nucleus → endosperm (triploid-3N)
egg nucleus + sperm
nucleus → zygote → embryo (diploid-2N)

Figures show the growth relationships among the different parts of a seed and fruit during development in an angiosperm plant. The embryo develops from the zygote initially as a microscopically small mass of cells embedded in the endosperm, which is in turn embedded in the nucellus. The embryo undergoes continuous morphological and physiological development. Three broad stages of development can be observed in embryos of most plants. For example, in *Capsella* and *Datura* in an early *proembryo* stage, the embryo is very small (microscopic) and globular in shape; it becomes *heart*-shaped as cotyledons begin to develop. This period is indicated by the 1- to 4-day lettuce embryo or the 10 to 12-day

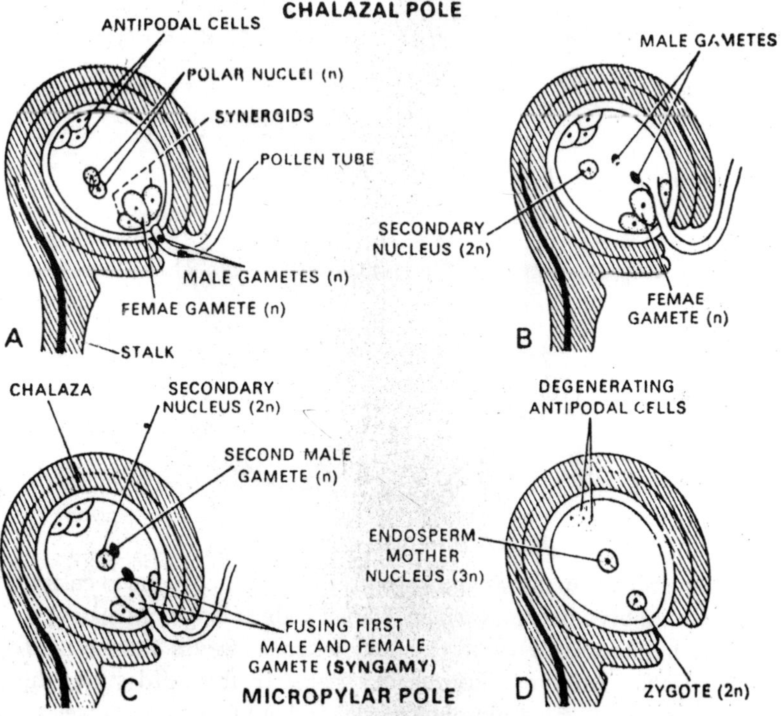

Fig. 4.1. L.S. of ovule showing different stages of fertilization.

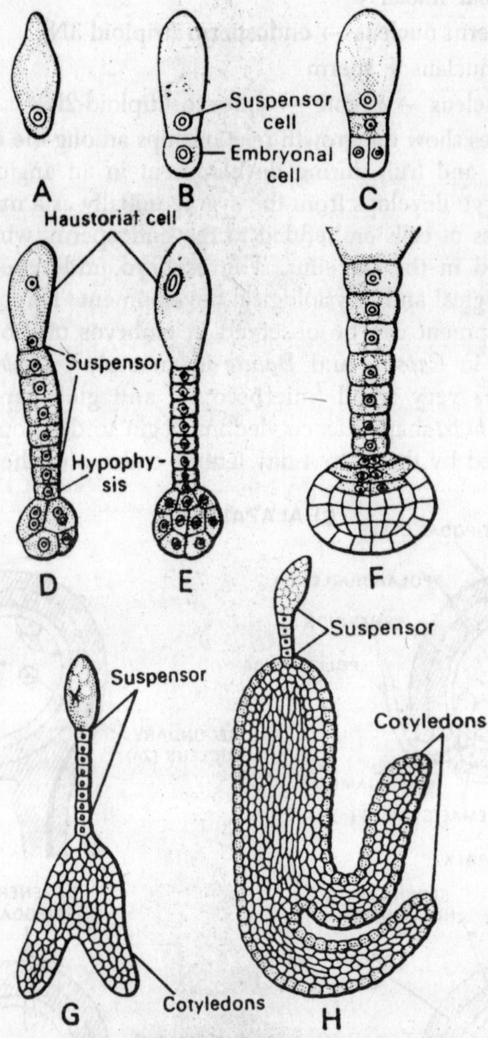

Fig. 4.2. Development of Dicot embryo.

Datura embryo. In the second stage the cotyledons continue to enlarge and the embryo becomes *torpedo*-shaped with its length increasing to nearly fill the seed. Five- to eight-day lettuce embryos and 15-to 20-days *Datura* embryos illustrate this stage. In the third stage, the embryos are full sized morphologically and are increasing in weight. The endosperm functions in a nutritional capacity for the embryo, although no vascular connections exist between the two structures.

Growth of the embryo is preceded by growth of the endosperm, which digests the inner nucellar tissue as it grows. The endosperm is, in turn, digested by the developing embryo, although in some species, a large part of the endosperm remains in the mature seed. Failure of the endosperm to develop properly results in retardation or arrest of embryo development. If embryo abortion result; the phenomenon is called *somatoplastic sterility*. It commonly occurs when two genetically different individuals are hybridized, either from two different species or from two individuals of different polyploid constitution. It, therefore, can be a barrier to hybridization. If such embryos are excised from the developing fruit before they abort, they can sometimes be saved by being grown in aseptic culture.

Seed development in the gymnosperms is somewhat different. The ovules are exposed within the ovular cone and are not enclosed in an ovary. The ovular cone develops into the mature cone. Gymnosperms do not undergo double fertilization. The embryo is nourished by the haploid female gametophyte and is not dependent upon a triploid endosperm. Consequently somatoplastic sterility is not a problem. Various external agents can prevent embryo development, even though the fruit itself continues to develop. Numerous kinds of insects attack the developing seed and fruit. This is particularly serious in forest trees. The developing seed of carrot and other Umbelliferous plants is attacked by *Lygus* bugs which can penetrate the fruit and feed on the embryo. Adverse weather, such as frosts during early fruit development, sometimes kills the embryo, but the fruit itself continues to develop. Growth tensions, causing the hardening endocarp to split or crack, can result in abortion of stone fruit seeds. This condition is associated with excessive vigor, reduced crop density, root restriction, or tree injuries resulting in girdling.

Accumulation of Food Reserves in the Seed

The accumulation of storage materials in the seed can be measured by changes in dry weight of the seed, although in the earlier part of the fruit development period, increase in weight may occur because of increase in size. Later, when the seed has attained its full size, the increase in dry weight is a measure of this accumulative process. These reserve materials originate as carbohydrate produced by photosynthesis in the leaves and translocated to the fruits and seeds where they are converted to complex storage products–carbohydrates, fats, and proteins. This process largely takes place in the final developmental periods of

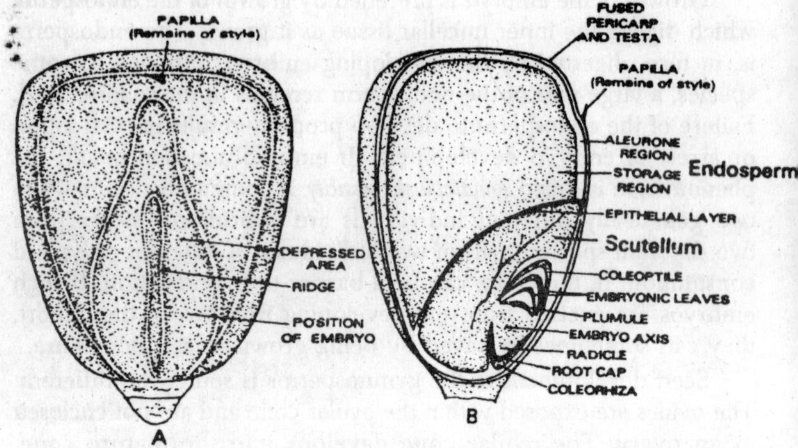

Fig. 4.3. Structure of Maize grain. A–External structure, B–L.S. or V.S. grain.

fruit growth. This accumulative process must take place properly if high-quality seeds are to be produced. Such seeds should be plump and heavy for their size. Since the initial growth of the seedling depends upon these reserve materials, heavier seeds should result in better germination and produce more vigorous seedlings. If conditions interfere with this storing process so that few reserve materials accumulate, the seeds will be thin, shrunken, and light in weight. The more severe this condition is, the less the seeds can survive storage periods, the poorer is the germination, and the weaker are the seedlings that are produced.

Ripening and Dissemination of Seeds

Specific physical and chemical changes take place during ripening which lead to the senescence of the fruit and the dissemination of the seed. One of the most obvious changes is the drying of the fruit tissues. In certain fruits, this leads to dehiscence and the discharge of the seeds from the fruit. Changes may take place in the colour of the fruit and the seed coats, and softening of the fruit may occur. An immature fruit is invariably green because of the presence of chlorophyll, but when the fruit ripens, the chlorophyll breaks down and may disappear, causing other colours to become evident as particular pigments are exposed.

The Mature Seed

Botanically, in the angiosperms the seed is a matured ovule enclosed within the ovary, or fruit. Seeds and fruits of different species vary greatly in appearance; size; shape; location and structure of the embryo; and presence of storage tissues. These points are useful in identification. From the standpoint of seed handling, it is not always possible to separate the fruit and seed, since they are sometimes joined in a single unit. In such cases, the fruit itself is treated as the "seed," as in wheat or corn.

Parts of the Seed

The seed has three basic parts : (*a*) embryo, (*b*) food storage tissues, and (*c*) seed coverings.

Embryo

The embryo is a new plant resulting from the union of a male and female gamete during fertilization. Its basic structure consists of an *axis* with growing points at each end, one for the shoot and one for the root, and one or more seed leaves (*cotyledons*) attached to the embryo axis. Plants are classified by the number of cotyledons. Monocotyledonous plants (such as the grasses or onion) have a single cotyledon, dicotyledonous plants (such as the bean or peach) have two, and gymnosperms (such as the pine or *Ginkgo*) may have as many as fifteen.

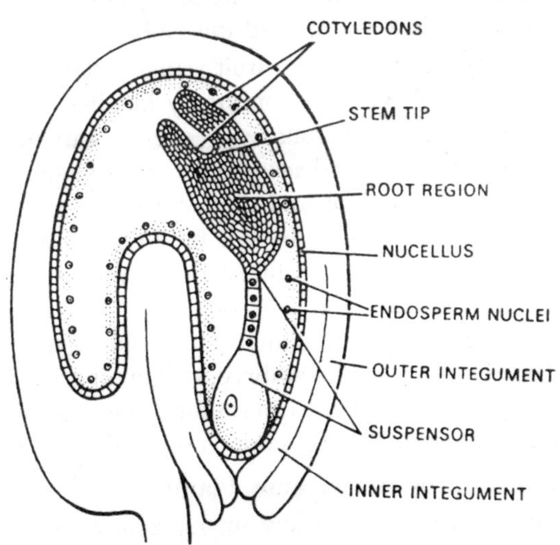

Fig. 4.4. L.S. of ovule showing young embryo.

Storage tissues

The storage tissues of the seed may be the cotyledons, the endosperm, the perisperm, or, in gymnosperms, the haploid female gametophyte. Seeds in which the endosperm is large and contains most of the stored food are referred to as *albuminous seeds*; seeds in which the endosperm is lacking or reduced to a thin layer surrounding the embryo are referred to as *exalbuminous* seeds. In the latter case, the stored food is usually within the cotyledons, the endosperm being digested by the embryo during development. The perisperm, originating from the nucellus, occurs in only a few families of plants, such as the Chenopodiaceae and the Caryophyllaceae. Usually it is digested by the developing endosperm during seed formation.

Seed coverings

The seed coverings may consist of the seed coats, the remains of the nucellus and endosperm, and sometimes parts of the fruit. The seed coats, or *testa*, usually one or two (rarely three) in number, are derived from the integuments of the ovule. During development the seed coats become modified, and at maturity they present a characteristic appearance. Usually the outer seed coat becomes dry, somewhat hardened and thickened, and brownish or otherwise coloured. On the other hand, the inner seed coat will usually be thin, transparent, and membranous. Remnants of the endosperm and nucellus are found within the inner seed coat, sometimes making a distinct, continuous layer around the embryo.

In some plants, parts of the fruit remain attached to the seed so that the fruit and seed are commonly handled together as the "seed." In certain kinds of fruits, *e.g.*, achenes, caryopsis, samaras, and schizocarps, the fruit and seed layers are contiguous. In others, such as the acorn, the fruit and seed coverings separate but the fruit covering is indehiscent. In still others, such as the "pit" or stone fruits or the shell of walnuts, the covering is a hardened portion of the pericarp, but it is dehiscent and can usually be removed without much difficulty. The seed coverings provide mechanical protection for the embryo, making it possible to handle seeds without injury, and thus permitting transportation for long distances and storage for long periods of time.

Seed Production of Herbaceous Plants

To keep a variety genetically unchanged, control of the seed source is necessary. With no control of the genetic characteristics of the variety may become somewhat different from, that initially developed by the plant breeder. Contamination may result from chance-pollination with plants of a different genotype, or it may

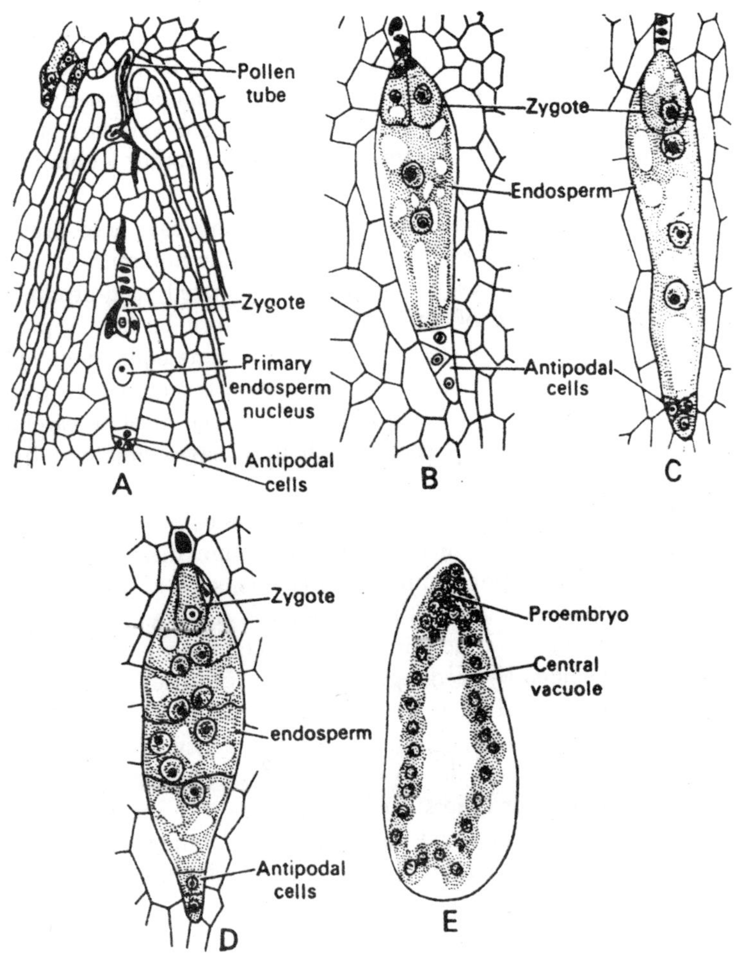

Fig. 4.5. Development of Endosperem–Nuclear type.

result from a mixture with other crop seed of lesser value or with weed seed. Foreign seed similar to the pure crop seed cannot be easily removed with seed-cleaning equipment.

Methods of Maintaining Purity

Isolation

Isolation is used (*a*) to prevent contamination by cross - pollination with a different but related variety, and (*b*) to prevent mechanical mixing of the seed during harvest. It is primarily achieved through distance, but it also can be attained by enclosing plants or groups of plants in cages, enclosing individual flowers, or removing male flower parts and then employing artificial pollination. Not as much isolation is needed in the production of self-pollinated plants as in the production of cross-pollinated plants. As a general rule, two or more closely related varieties of self-pollinated plants can be planted close to each other without one contaminating the other.

The principal reason for separation in this case is to prevent mechanical mixing of seed during harvest. The minimum distance usually specified between plots of self-pollinated seed-producing plants is 10 ft. Seed of different varieties must be kept separate during harvest. Careful cleaning of the harvesting equipment when a change is made from one variety to another is required. Likewise, sacks and other containers used to hold the seed must be carefully cleaned to remove any seed which may have remained from previous lots.

Isolation is a primary factor in the production of plants cross-pollinated by wind or insects. The minimum distance between varieties depends upon a number of factors, such as the degree of natural cross-pollination, the pollination agency, direction of prevailing winds, and the number of insects present. The required distances specified by seed-regulating agencies varies with the class of seed produced. That is, greater distances should be used with seeds in which a high degree of genetic purity is desired. The minimum distance usually specified for insect-pollinated plants is ¼ mi., although for others, such as the onion, 1 mi, is specified. Some seed producers have indicated that a minimum of ½ mi, should be used, preferably 1 mi, or more, depending upon the location of colonies of bees.

The distance that should separate varieties of wind-pollinated plants varies with different kinds of plants. The distance usually specified for corn is 40 rd (½ mi). In hybrid seed corn, this distance may be reduced by planting border rows of the pollinating variety. With beets, ½ to 1 mi, is recommended, and to produce stock seed 2 mi, is preferable. Seed-producing fields of two different varieties should not be in the line of prevailing winds.

Cross-pollination takes place between certain varieties and not between others. In general, any variety can contaminate any other variety of the same species, it may or may not contaminate varieties of a different species but in the same genus; and rarely will it contaminate varieties belonging to another genus. Since the horticultural classification may not indicate taxonomic relationships, the seed producer should have some familiarity with botanical relationships among the varieties he grows.

Roguing

The existence of off-type plants, plants of other varieties, and weeds in the seed production field are other sources of contamination which should be eliminated. Although a low percentage of such plants may not seriously affect the performance of any one lot of seed, their continued presence will lead to deterioration of the variety over a period of time. The removal of such plants is referred to as "roguing".

Off-type plants may arise because some recessive genes are present in a heterozygous condition even in highly homozygous varieties. Recessive genes may arise by mutation, a process continually occurring at a low rate. The effect of a mutant recessive gene controlling a given plant characteristic may not be immediately observed in the plant in which it arises. The plant becomes heterozygous for that gene, and in a later generation the gene segregates and the character appears in the offspring. Off-type individual plants should be rogued out of the seed production fields before pollination occurs. Regular supervision of the fields, by trained personnel is required. Volunteer plants arising from accidentally planted seed or from seed produced by earlier crops is another source of contamination. Fields for producing seed of a particular variety should not have grown a potentially contaminating variety for a number of preceding years.

Testing

Varieties being grown for seed production should be periodically tested for genetic purity to make sure that they are being maintained in their true form. Many seed companies maintain test gardens for this purpose under the supervision of trained personnel.

Genetic Shifts

Sometimes a seed-propagated variety that is produced initially in one given environment may undergo a *genetic shift* if it is grown for several consecutive generations in a different environment. Such a change takes place because those seedling plants better adapted to the new environment survive in greater amounts than others. Physiological changes affecting yielding ability, disease resistance, and environmental adaptation are difficult to detect without testing. To minimize the opportunity for such shifts to occur in seeds of forage crops, only one generation of seed production is allowed in regions of mild winters if the crop seed is to be used for planting in regions with cold winters.

Seed Certification

Genetic purity in commercial seed production is often maintained through a system of *seed certification.* Such programs exist in most states of the U.S through the cooperative efforts of public research, extension, and regulatory agencies in agriculture and a state Crop Improvement Association. The Crop Improvement Association is usually a legally designated seed-certifying agency. Its members include growers as well as other interested individuals, involved in the production of "certified seed." These individual state organizations are coordinated through the International Crop Improvement Association with members in the United States and Canada.

The principal objectives of seed certification are to maintain and to make available crop seeds, tubers, or bulbs and sometimes turfgrasses, which are of good seedling value and true to variety. To accomplish these purposes, the seed certifying agency determines the eligibility of particular varieties. It sets up production standards for isolation, presence of off-type plants, and quality of harvested seed, and makes regular inspections of the production fields to see that the standards are maintained. Supervision of seed processing is also involved. The International Crop Improvement Association has recommended minimum standard for many crops, although these regulations are set within the individual state organization. Since specific requirements for certification may vary from state to state, the local state regulations should be examined. These usually specify the varieties which can be certified, the amount of isolation required of any particular crop, required inspections, and standards for the seed quality after harvest.

The mechanics of maintaining genetic purity are facilitated by utilizing certain classes of seed which are maintained at different levels of purity. These classes are given below, as defined by the International Crop Improvement Association.

Breeder's seed

Breeder's seed is seed or vegetative propagating material which is directly controlled by the originating or, in certain cases, the sponsoring plant breeder or institution, and which provides for the initial and recurring increase of foundation seed.

Foundation seed

Foundation seed, including *elite* in Canada, is seed stock so handled as to most nearly maintain specific genetic identity and purity, and that may be designated by an agricultural experiment station. Production must be carefully supervised or approved by representatives of the station. Foundation seed is the source of all other certified seed classes, either directly or through registered seed.

Registered seed

Registered seed is the progeny of foundation or registered seed that is so handled as to maintain satisfactory genetic identity and purity, and that has been approved and certified by the certifying agency. This class of seed should be of a quality suitable for the production of certified seed.

Certified seed

Certified seed is the progeny of foundation, registered, or certified seed that is so handled as to maintain satisfactory genetic identity and purity, and that has been approved and certified by the certifying agency.

Certified seed is the class grown in largest volume to be placed on the market and sold to growers. Bags of certified seed have attached, with a metal seal, a blue tag distributed by the seed-certifying agency as evidence of the genetic identity and purity of the seed contained therein. Registered seed is used to produce the plants from which certified seed is to be harvested. Bags of registered seed are labeled with a purple tag or a blue tag marked with the work "registered." Like-wise, foundation seed is labeled with a white tag or a regular certified seed tag with the word "foundation."

Vegetable and Flower Seed Production

In the U.S. vegetable and flower seed is largely produced by commercial seed companies who either grow the seed themselves or contract with private growers to produce seed for them. The company preserves genetic purity by specifying the particular variety an individual contractor will grow, by supervising the fields during production and by maintaining test gardens.

Although such seed is not necessarily produced as certified seed, similar procedures are used to maintain genetic purity. Essentially the same classes of seed are used, although they may be designated by different names. The primary source of seed of the variety, the foundation seed, is usually maintained by the seed company. Trained personnel have the responsibility for maintaining the foundation seed at the highest level of purity. Seed provided by the company to the contractor-grower for growing his crop is known as *stock seed.* The seed actually harvested from the fields of the contractor-grower will be placed on the market and sold commercially by the seed company.

The Production of Seed for Hybrid Varieties

Hybrid varieties have become an increasingly important category of cultivated plants within recent years. These are the progeny produced by the repetitive crossing of two or more parental lines that are maintained either (*a*) by seed, such as inbred lines, or (*b*) asexually, such as clones. To produce commercial hybrid seed, the parental lines must be grown side by side so that cross-pollination takes place between them. The seed produced (the F_1 progeny of the cross) is the seed used to grow commercial crops. This cross must be repeated every time the seeds are produced.

Hybridization between two inbred lines is known as a *single-cross.* Combining two single crosses produces a *double-cross,* the usual case with hybrid corn varieties. Isolation of the seed-producing fields should be even more stringently observed than in other cross-pollinated crops. With enforced cross-pollination, the possibility of pollen contamination is greater. Hybrid plants were first utilized in the production of field corn, and almost all the field corn grown in the United States is of this type. Likewise, hybrid varieties of sweetcorn and popcorn have been developed. Hybrid varieties of other plants have not been developed as rapidly as those of corn, the limiting factor in most cases being the development of methods whereby the hybrid variety can be produced on a large scale at a moderate cost. Nevertheless, a number of vegetable hybrids have become available as have hybrids of some flower plants (for instance, petunia).

Methods of Cross-pollination

(a) Plants of monoecious species (where male and female organs are in different flowers) readily adapt to hybrid seed production. To produce hybrid corn seed, the parental lines are planted in rows, one male row to three female rows or a similar combination. Production of pollen is not allowed to take place in seed-producing rows, and thus pollen to fertilize the female flowers (the silks) will come only from the male parental lines. This is done either by removing the male flowers (tassels) by hand or by using pollen-sterile plants as described later.

(b) Controlled hand pollination is sometimes practiced in producing seeds from particularly selected parents. Although used mostly in plant breeding, it is also used to produce some commercial hybrid seeds. The specific technique used depends upon flower structure and the pollination characteristics of the individual species, but it follows certain principles. Accidental contamination by unwanted pollen can be prevented by washing the hands and instruments with alcohol before handling pollen or flowers. Pollen should be collected from flower buds immediately preceding bloom and before the anthers open (dehisce). This eliminates the chance of contamination by unwanted pollen.

Anthers are extracted by pulling them from the flowers with tweezets, by squeezing the flower between the fingers, or by rubbing the buds across a wire screen. Catkins (walnut, birch, aspen) or staminate cones(pine) will open on drying and will shed large quantitics of pure, dry pollen if placed on a sheet of paper in a warm room. Anthers are dried on a sheet of paper until their opening can be observed under a magnifying glass. Pollen and anthers can be screened through a fine sieve to remove extraneous material; they are then stored in a glass vial or bottle.

Most kinds of pollen will remain viable for only a few days or weeks at warm temperatures, but many kinds can be preserved for several months to several years if stored at low temperatures and relatively low humidity. Effective storage conditions include a combination of 10 to 50 percent relative humidity and a temperature of 32° to 50°F (0° to 10°C). Moisture content of the pollen can be controlled by storing over a desiccant, such as calcium chloride or sulfuric acid. Some pollen–that of grasses, for instance–is better stored at 90 to 100 per cent R.H. Pollen can be relatively stored at about 0°F (–18 C), as in a home freezer.

The stigma at the tip of the pistil must be exposed to apply the pollen. If the plant is self-fertile, the stamens must be removed (emasculated). Petals and stamens can be removed at the bud stage either with tweezers or by cutting the base of the flower between the fingernails. If the flower is to be self-fertilized or is a self-sterile variety, the flower need not be emasculated but only enclosed to keep out pollinating insects or wind-blown pollen.

Pollen is applied to the sticky, receptive stigma of insect-pollinated flowers with a fine-hair brush, a glass rod, a pencil eraser, or the tip of the finger. Removal of all the flower parts with the exception of the pistil reduces or eliminates the likelihood of further chance pollination by insects. Complete protection can be obtained by covering the individual flower, the pistil, or the branch with cellophane, plastic, paper, or a cloth bag.

For wind-pollinated plants, *e.g.*, grasses, pines, and walnut, female flowers (or cones) must be carefully covered throughout the flowering period. They can be covered with cellophane, paper, or a tightly woven cloth bag. Pollen can be transferred by inverting a paper bag containing pollen over the exposed [illegible] or dry pollen can be blown into the bag with an atomizer or inserted with a hypodermic needle. The flower or branch on which it is borne should be carefully labeled to maintain a record of the pollen used. The usual method is to identify the cross as follows: "seed parent × pollen parent."

(c) Mass production of hybrid seed of most species is possible only by utilizing or developing some system of self-sterility the prevents self-fertilization. Enforced hybridization results when a self-sterile, seed-producing line is grown with an appropriate pollen-producing parental line and some vector for cross-pollination is present. The major difficulties in utilizing this procedure in commercial seed production are (1) discovering suitable self-sterility genes, (2) incorporating them into useful varieties (a job of plant breeders), and (3) maintaining the necessary self-sterile genotype in the perental lines. If parental lines can be economically propagated vegetatively (or by apomixis) then self-sterility can be maintained intact with no difficulty. Seed-propogated lines, however, are more difficult to manage and a thorough knowledge of inheritance in the self-sterility system is required. Methods must be used to insure that only the desired self-sterile genotype is produced when needed or procedures must be used to identify the unwanted pollen-producing individual plants early enough in their development so that they can be rogued out of the seed-producing fields.

Actually, complete cross-pollination might not be necessary. Burton describes a system utilized in the hybrid-seed production of 'Gahi-1', a partially cross-pollinated variety of pearl millet (*Pennisetum glaucum*) used for forage. Two or four selected parental lines are grown together, and the seed is harvested from the entire block and mixed thoroughly. About 75 per cent of the offspring are hybrids. When planted at high density, the more vigorous hybrid seedlings crowd out the weaker inbred seedlings. At lower planting densities, mixtures of hybrid and inbred plants result. Later in this chapter, a similar procedure is described for some tree hybrids.

Several naturally occuring systems of enforced cross-pollination exist and are utilized in the seed production of some hybrids. Self-incompatibility has been used in the production of some cabbage hybrids. Some grass hybrids of Pensacola Bahiagrass (*Paspalum notatum*), for instance, have been produced by interplanting self-incompatible (but cross-compatible) parental clones. The plants spread vegetatively to occupy the field and all seeds are harvested together as hybrids. Dioecious plants can also be utilized effectively in this way. 'Mesa' buffalograss (*Buchloe dactyloides*) is produced by interplanting vegetatively propagated male and female parents in isolated fields.

Genes for male sterility that inhibit pollen production have been discovered in many crop plants and have been incorporated in the breeding lines of a number of important crop species. Male sterility is inherited in several ways in different species and not all systems are effectively used in producing hybrids. The most significant one is the cytoplasmic-genetic male-sterility system which was first utilized for the production of onion hybrids. It is also used in the commercial production of hybrid seed of field corn, sugar beets, sorghum, and pearl millet. Since it can be used for other crop plants as well, its use will undoubtedly be extended. The system involves the combination of a cytoplasmic factor S (as opposed to a normal cytoplasm N) and a recessive gene ms(as opposed to a dominant fertility-restoring gene Ms). A male-sterile plant has a genotype of S ms ms, whereas a male-fertile plant can have any of the following : N ms ms, N Ms ms, S Ms Ms, or S Ms ms. The male-sterile line is maintained by continuous back crossing to a pollen parent of N ms ms. Two examples of the inheritance of these factors are the following :

(*a*) S ms ms × N ms ms
(seed) (pollen)
↓
S ms ms
(all male-sterile)

(*b*) S ms ms × N ms ms
(seed) (pollen)
↓
½ S Ms ms
(male-fertile)
½ S ms ms
(male -sterile)

Use of F_1 Seed for Propagation

From the standpoint of continued propagation, one important point should be emphasized in regard to hybrid varieties. The F_1 hybrid plants are heterozygous, and seeds from them will not continue to produce plants with the characteristics of the F_1 plants. The effect of hybrid vigor is greatest in the F_1 generation and declines sharply in succeeding generations. Such seed would be unreliable for propogation purposes. For each crop that is to be grown, new hybrid seed must be produced according to the procedure described. The crop grower must rely upon the seeds man to provide him with an adequate supply of properly produced seed.

Seed Sources for Woody Perennials

Most trees and shrubs are heterozygous and cross-pollinated and have considerable potentiality for genetic variability. Yet there are important reasons for growing tree seedlings. Seedling production involves two problems : variability and the unpredictability of the ability of many seed sources to produce trees of known characteristics. Growing seedlings of woody plants directly involves the propagator in the process of effecting improvement in tree crops. Woody plants propagated from seed are used for various purposes, such as landscaping, rootstocks (fruit trees and ornamentals), Christmas tree growing, and reforestation. The amount of allowable variation within a group of plants of seedling origin depends upon their intended use. In growing trees and shrubs for uses in which the ultimate characteristics of the individual plant should be known, such as landscaping or Christmas tree growing, either careful seed selection is required or vegetative propagation methods are needed. Uniformity is important among root-stock plants and is achieved in producing fruit tree rootstocks by preserving the seed source as a clone. In some reforestation practices a certain amount of seedling variation may be beneficial since initial planting density is usually high. Close spacing promotes good stem form, and competition among plants gradually eliminates the weakest trees. On the other hand, modern forestry practices are more and more involved with the growing of carefully selected trees of good timber quality.

PRINCIPLES OF SELECTION

Seed Origin

Seed origin refers to the particular climatic and geographical locality where the seed is obtained. The original geographic source of seed is referred to as *provenance.* It is particularly important in reforestation and is second in importance only to choice of species. The significance of seed origin was first recognized during the nineteenth century when extensive plantings of forest trees from unselected seed were made in Europe. Observers notices that trees grown from imported seeds were generally inferior to trees grown from seeds of local origin. In one instance, thousand of acres of Scotch pine (*Pinus sylvestris*) were eradicated because of their inferior characteristics. After that many European countries enacted laws covering the importation and use of forest tree seeds, in part requiring the labeling of seed as to origin.

Origin is important when plants of single species grow over a wide range of ecologically different areas in nature. Variation in morphology, physiology, adaptation to climate and soil, and resistance to diseases and insects may exist the represent specific races, or *ecotypes*, that are characteristic of a particular locality. Variation may also exist in a continuous gradation, or *cline.* A locality may be defined by its latitude, longitude, and elevation. Seeds of a given species collected from one locality may produce plants completely unadapted to a different locality. For example, seed collected from trees in warm climates or at low altitudes is likely to produce seedlings that will be injured when grown in colder regions even if the species is the same and the plants are similar in appearance. Although the reverse situation–collecting seed from colder areas for growth in warmer regions–would be more satisfactory, it might result in a net reduction in growth due to the inability of the trees to fully utilize the growing season.

Distinct racial variation has been demonstrated to occur in the United States in Douglas fir, ponderosa pine, lodgepole pine, red pine, eastern white pine, slash pine, loblolly pine, shortleaf pine, and white spruce. Evidence shows that similar variations occur in other native tree crops, although details are lacking. On the other hand, some seed sources have been shown to be superior to others and preferable to most local sources–for example, the East Baltic race of Scotch pine; the Hartz Mountain source of Norway spruce, the Sudeten strain of European larch, the Burmese race of teak, Douglas fir from the Palmer area in Oregon, ponderosa pine from the Lolo Mountains in Montana, and white spruce from the Pembroke, Ontario area.

An important principle in seed origin is to use local seed where possible unless adequate research has demonstrated that another source is better. In 1939 the United States Department of Agriculture adopted the following policy regarding the use of seed stocks and individual clones obtained by them for forest, shelter-belt, and erosion-control plantings. It serves as a guide for the selection of woody plant seed from natural sources.

1. Only seed of known locality of origin or nursery stock grown from such seed is to be used.
2. Evidence for the place and year of origin is to be required of the vendor when seeds are bought.
3. An accurate record of the following data is to be required of all shipments : (*a*) lot number; (*b*) year of seed crop; (*c*) species; (*d*) seed origin as to state, county, locality, and range of elevation; and (*e*) proof of origin.
4. Whenever available, local seed from natural stands is to be used unless it is demonstrated that another source can give desirable plants. "Local seed" means seed from an area subject to similar climatic influences, and this may usually be considered to mean within 100 mi, of the planting and within 1000ft of its elevation.
5. When local seed is not available, seed should be used from a region having as nearly as possible the same climatic characteristics, such as length of growing season, mean temperature of growing season, frequency of summer droughts, and latitude.

Seed-collection zones designating particular areas that have a specific climatic and geographical basis have been established in a number of area in the U.S. For example, California has 14 such zones and 5 additional sub-zones. Similar zones are established in Washington and Oregon and in the central states are of the United States.

Heit has shown that choice of seed sources is highly important in growing conifers and various hardwood species of trees in the nursery–for instance, for landscape purposes or for Christmas trees. These differences were demonstrated in nursery test plants. For example, in his tests Douglas fir had at least three recognized races, *viridis*, *caesia*, and *glauca*, with various geographical strains within them. The *viridis* strain from the West Coast was not winter-hardy; strains collected further inland were winter-hardy and vigorous, growing. Trees of the *glauce* (blue) strain from the Rocky Mountain region were winter-har , but varied in growth rate and appearance. Similarly, differences among sources occurred in Scotch pine, mugho pine, Norway spruce, and others.

Seed-Tree Selection

Although local sources of seed of forest species can produce the best adapted plants for a given site, selection of *individual seed trees* may be necessary to improve quality in other characteristics. In forest trees there is a great deal of evidence that the quality of the individual seed trees is a good indix of the quality of its offspring. The more knowledge the collector has about the transmission of specific characteristics from parent to offspring, the more accurate selection will be. Parent tree selection for many characteristics, such as stem form, branching habit, growth rate, resistance to diseases and insects, presence of surface defects, and certain other lumber qualities, can be effective.

Selected trees with a superior phenotype are referred to by foresters as "plus" trees and are often not cut for lumber. Such individual trees have value for their seeds either for natural reseeding or as seed sources for other areas. They can be multiplied and maintained as a clone in an established seed orchard.

Seeds of many woody plants used in nurseries are collected wherever the species is found growing locally, as in orchards, forests, wood lots, streets, and roadsides. A particular seed source should be selected not only for convenience of collection but, more important, because seed obtained from it possesses desirable genetic potentialities. Consequently an initial step in selecting a seed source is to evaluate the characteristics of the seed trees themselves. Although plants possessing the desired characteristics are most likely to pass them on to their offspring, this assumption cannot be safely made without progeny testing.

A second step in selecting a seed source is to evaluate the characteristics of surrounding trees that could cross-pollinate the seed trees. Most woody tree species are cross-pollinated ; in general, better quality seeds will likely result thereby. Consequently it is undesirable to collect seed from an isolated plant or from those growing near others of related species, such as they might be in an arboretum. Cross-pollination from unwanted sources can nullify the potential value of a given seed source by producing excessive and unpredictable variability. For example, prior to World War I, "Old French" pear seedlings were grown from seed obtained in Europe from *Pyrus communis* trees. Much variability resulted from hybridization with *Pyrus nivalis.* Similarly, the "Oriental pear rootstock" species obtained during that same period were from seeds obtained from Asia with little knowledge about the characteristics of either the seed tree or the pollen source. Plants grown from these sources have turned out to be a heterogeneous group with many species and hybrids. The differences among the flame eucalyptus (*Eucalyptus ficifolia*) plants arising from separate collection of a particular source in Australia have been explained by the fact that seeds collected from the center of the block were likely to breed true and produce upright, uniform groups of trees those from the edge of the block were likely to be hybrids with surrounding *E. calophylla* trees, and consequently produced many dwarfed and off-type plants.

The most desirable procedure is to collect seeds from groups of the same kind of plants with desired characteristics growing in a *pure stand.* Seedling plants grown from such sources, although not necessarily homozygous, would be most likely to reproduce the characteristics observed among the parent seed trees.

Control of both seed and pollination source can be achieved by collecting from trees of clonal varieties growing in orchards with a known pollination arrangement. Such a procedure has been used extensively to produce rootstock seeds in fruit growing. For example, to obtain true *Pyrus communis* seedlings *e.g.,* the so-called "French pear", seeds can be obtained from trees of 'Winter Nelis' growing as scattered trees for cross-pollination in'Bartlett' ('Williams Bon Cretien') pear orchards. All seedlings would be 'Winter Nelis' × 'Bartlett' hybrids. Likewise, 'Lovell' peach seeds or 'Royal' apricot seeds collected from fruit-drying yards come from known clones that most likely are grown in a solidly planted block of trees where self-pollination occurred.

Progeny Testing

A true genetic value of a seed source can only be established through a *progeny test.* A representative sample of seeds is planted, and the resulting progeny are grown under test conditions that will identify essential characteristics or demonstrate presumed superiority to other sources, or both. Foresters refer to seed-source trees with a superior genotype, as demonstrated by a progeny test as "elite" trees.

Progeny testing has been used to identify fruit-tree clones that will transmit high yielding characteristics–as in tung (*Aleurites*)–

or important rootstock qualities, such as nematode resistance. Heit has used nursery tests to identify and characterize specific seed sources for landscape and Christmas tree uses. Progeny testing is an essential phase of identifying seed sources of superior "elite" forest trees. Here the procedure is particularly difficult, since the tester is working with a wild species with no standard cultivars for comparison requiring a long-term evaluation of perhaps 20 years or more.

Nursery Selection

Desired individuals can be obtained and variability reduced to some extent through *nursery selection.* This procedure requires that variability involve identifiable characters of vigor or appearance, or both. Paradox hybrid walnut seedlings can be identified in nursery plantings of *Juglans hindsii* seedlings. "Blue" seedlings of the Colorado spruce (*Picea pungens*) appear among others having the usual green form. Differences in fall colouring of *Lequidamber* and *Pistacia* trees among seedlings necessitates fall selection of individual trees for landscaping. Variation among seedlings grown for rootstocks may be reduced by grading out the week and small seedlings, a usual practice in nurseries in the United States.

Procedures in Selection of Woody Plant Seed

Seed-Collection Areas

A *seed-collection* area is one where the trees have promise as a potential seed source have not been progeny-tested. This should be a pure stand of uniform trees large enough to insure adequate cross- pollination and separated from other planıs having undesirable characteristics., Such areas are not manages or treated specifically for seed production.

Seed-Production Area

This refers to an area enclosing a group of trees that has been set aside permanently as a seed source. Use of such an area for a seed source applies primarily to forest trees, but the procedure could be utilized in any woody plant seed enterprise. Seed trees within the area are evaluated for the characteristics desired. Off-type trees or those that do not meet minimum standards are rogued out and eliminated. Other trees or shrub species that would interfere with the operations may be removed. It may be desirable to remove additional trees to provide adequate space for tree development and seed production. An isolation zone at least 400ft wide from which off type trees are removed should be established around the area. Trees may or may not have been progeny tested prior to establishment of the area.

Seed Orchards

These are orchards or plantations established specifically for seed production, usually from seed trees of acceptable origin and quality, and preferably from progeny- tested seed trees. Seed orchards are used directly or indirectly in the production of fruit-tree rootstock seeds and to a limited extent in the production of forest trees. This procedure will undoubtedly increase as progeny-tested seed sources become available. Seed orchards are useful for maintaining rootstock seed trees involved in virus- control programs.

There are three general types of seed orchards. They may be made up of (a) seedlings produced from selected parents through natural or controlled pollination; (b) clonal seed orchards in which selected clones are propagated by grafting, budding, or rooting cuttings and grown in the block; and (c) seedling-clonal seed orchards in which certain clones are grafted into branches of some of the trees. The choice of which type should be used will depend upon the species that is being grown.

Details of setting up a seed orchard vary with the species. A site should be selected for maximum, efficient production. For most native species enough different tree selections should be included in a suitable arrangement to insure cross- pollination and to decrease effects of possible inbreeding. If the species is dioecious, pollen-producing trees must be included. Fruit-tree rootstock clones that are self-pollinated can be planted as solid blocks. An isolation zone at least 400ft wide should be established around the orchard. The size of this zone may be reduced if a buffer area of the same kind of tree is established around the orchard.

Hybrid Seed Production

Production of F_1 hybrid seeds is used to grow some tree crops. The principles are the same as those described earlier in the chapter. These seeds result from crosses between species or within species. Hybridization between the parental plants to produce the desired F_1 offspring may be achieved by various means. (a) Hybrids may be produced by hand pollination. This procedure is used mostly to produce F_1 hybrids for testing purposes. Mass production of seeds by this method would normally be too expensive, although hybrids of *Pinus rigida* × *P. taeda* have been produced this way in Korea.

(b) Hybrids may be produced in seed orchards by interplanting the parental trees (1). Hybrids of *Larix decidua* and *L. leptolepsis* have been produced in seed orchards in Europe. In Idaho, a western white pine (*Pinus monticola*) seed orchard was established in the late 1950's containing 13 parental clones selected by progeny testing. Their F_1 offspring showed an average of 30 percent resistance to blister rust. The F_2 generation showed an even higher average level of resistance in tests and will eventually be utilized as a seed source. (c)Another method is illustrated by the production of Paradox hybrid walnut seed, a procedure probably limited to certain kinds of plants. Such seed is obtained by collecting from individual Northern California black walnut (*Juglans hindsii*) seed trees where natural cross-pollination is known to take place with pollen from nearby English walnut (*Juglans hindsii*) seed trees where natural cross-pollination is known to take place with pollen from near by English walnut (*Juglans regia*) trees. Seed-source trees are identified by progeny testing. Individual Paradox seedlings are identified in the nursery by vigor and leaf characteristics. (d) Seed from F_1 seed trees gives F_2 or second-generation hybrids. Collection of seed from many such sources, particularly if interspecific hybrids are involved, produces a group of seedlings that is highly variable in regard to vigor and other characteristics.

As a group these may fail to reproduce the desirable attributes of the F_1 hybrid generation, although some individuals may be highly desirable. Such variability is undesirable if the hybrid plants are to be used for rootstocks or for landscaping were uniformity and predictability of the final form is important. On the other hand, in forest trees, some variability in vigor commonly associated with second-generation hybrids may have advantages. In tree plants with relatively high initial seedling density, competition among plants will favour the most vigorous and tend to eliminate the weaker, less desirable plants. Consequently, planting seeds from F_1 hybrids (the F_2) may be an economical way to obtain a natural stand of vigorous hybrids, F_2 hybrids of selected lines within species may be quite uniform and desirable as in the blister-rust-resistant, white pine hybrids described in an earlier paragraph. The value of an F_1 hybrid tree as a seed source should be established by progeny testing, the same as should that of any other seed source.

5

SEED GERMINATION

A seed consists of an embryo and its stored food supply, surrounded by protective seed coverings. When the seed separates from the plant on which it was produced, the seed is *quiescent*; that is, there is no external evidence of activity within the seed. The resumption of active growth by the embryo, resulting in the rupture of seed coverings and the emergence of a new seedling plant capable of independent existence, is known as *germination*. In order for germination to occur, three conditions must be fulfilled. *First* the seed must be *viable*; that is, the embryo must be alive and capable of germination. *Second*, *internal conditions* within the seed must be favourable for germination; that is, any physical or chemical barriers to germination must have disappeared. *Third*, the seed must be subjected to favourable *environmental* conditions, the essential factors being available water, proper temperature, a supply of oxygen and sometimes light. Although in any one seed each of these conditions may have an effect distinct from the others, the beginning of germination may be more often determined by the interactions among them.

The Germination Process

The germination process involves a complex sequence of biochemical, physiological, and morphological changes in which certain stages can be recognized. The *first* stage begins with the inhibition of water by the dry seed, the softening of the seed coverings, and the hydration of the protoplasm. This process is largely physical and occurs even in a non-viable seed. As a result of water absorption the seed swells, and the seed coats may break.

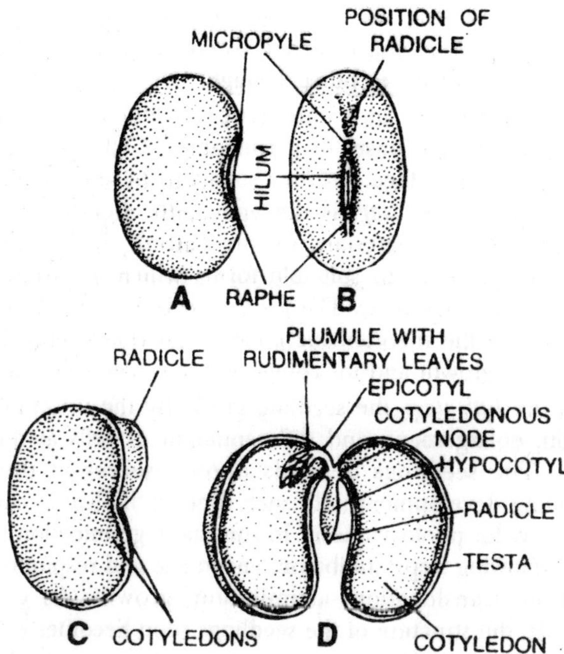

Fig. 5.1. Structure of Bean seed. A-B–external structure. C–Seed coats removed. D–Seed cut open to show various parts.

The *second stage* begins with the initiation of cellular activity and includes the appearance of specific enzymes and a rise in respiratory rate. A plant growth hormone–gibberellin–is believed to play a key role in the initiation of germination. This effect has been demonstrated most clearly in seeds of cereal grains, such as wheat and barley. In these seeds the important structures are the *embryo*, *endosperm* (nonliving storage tissue containing mostly starch), and the *aleurone* (the outer layer of endosperm one or two cells thick). When the dry seed imbibes moisture, gibberellin appears in the embryo and is translocated to the aleurone layer where it activates

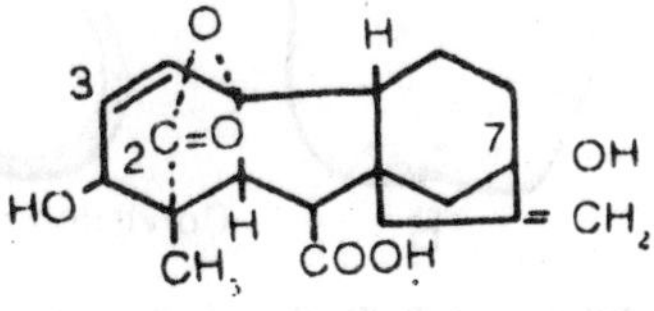

Fig. 5.2. Chemical structure of gibberllic acid (GA_3), a plant growth hormone associated with seed germination.

enzymes. On of these enzymes, alpha-amylase, moves to the endosperm, causing the starch to be converted to sugar. Other enzymes appear in the aleurone, weakening the seed coats and allowing the root tip to burst through.

Cell elongation and emergence of the radicle are events associated with the onset of germination. Cell division may also occur at an early stage, but this seems to be independent of cell elongation. A *third stage* is the enzymatic digestion of complex insoluble reserve materials, mostly carbohydrates and fats but sometimes proteins, to soluable forms which are translocated to the active growing areas. The *fourth stage* is the assimilation of these substances at the meristematic areas to provide energy for cellular activity and growth and for conversion into new cell components.

In the *fifth stage,* the seedling grows by the usual processes of division, enlargement, and differentiation of cells at the growing points. The seedlings depends upon reserves in the seed for continuing development until such time as the leaves can function adequately for photosynthesis. In summary, germination takes place in the following steps: inhibition, enzymatic and respiratory activity, digestion, translocation, assimilation, growth. As germination proceeds, the structure of the seedlings soon becomes evident. The embryo consists of an axis bearing one or more seed leaves, or *cotyledons.* The growing point of the root, the *radicle*, emerges from the base of the embryo axis. The growing point of the shoot, the *plumule,* is at the other end of the embryo axis, above the cotyledons. The seedling stem is divided into the section below the cotyledons–the *hypocotyl*–and the section above the cotyledons–the *epicotyl.* The initial growth of the seedling follows two patterns. In one type–*epigeous* germination–the hypocotyl elongates and raises the cotyledons above the ground. In the other type–*hypogeous*

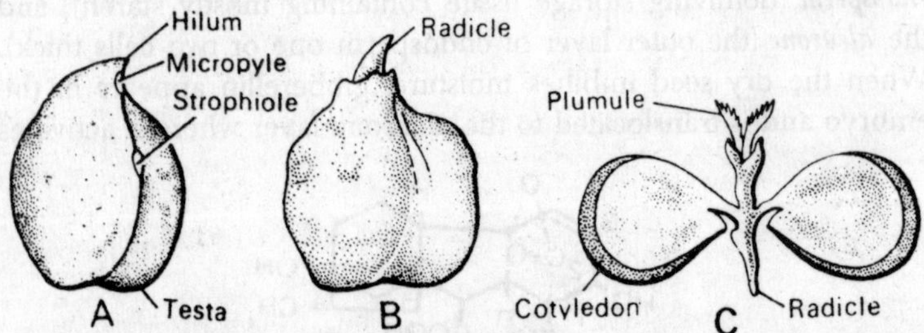

Fig. 5.3. Germ seed: A–Whole seed, B–The kernel after removal of seed coat, C–The two opened cotyledons with tigellum.

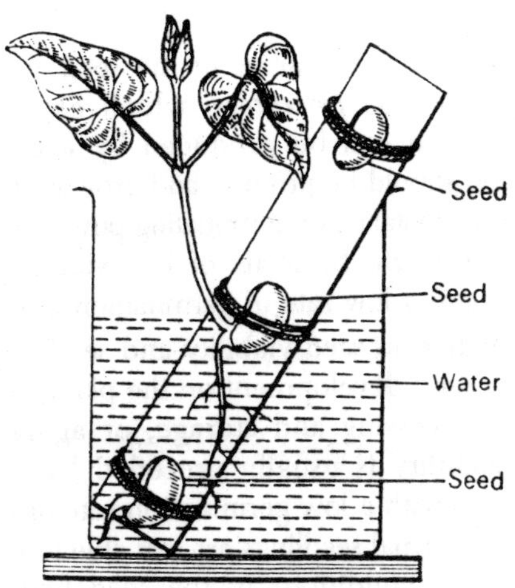

Fig. 5.4. Experiment to demonstrate conditions necessary for germination of seeds.

germination–the lengthening of the hypocotyl does not raise the cotyledons above the ground, and only the epicotyl emerges.

Viability of Seeds

A supply of viable seed is essential in successful propagation from seed. However, the distinction between a live and a dead seed is not necessarily sharp, and is often characterized by necrosis

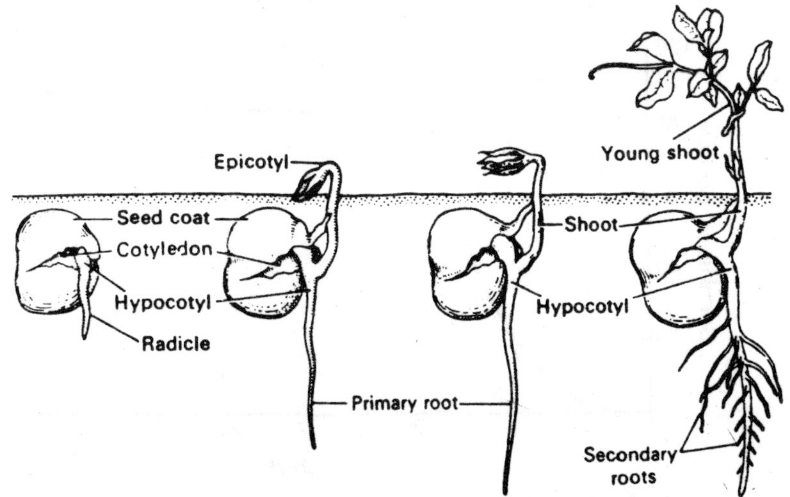

Fig. 5.5. Successive stages of hypogeal germination of dicotyledonous exalbuminous seed of pea.

or injuries in localized areas of the seed. Viability is represented by the *germination percentage*, which expresses the number of seedlings which can be produced by a given number of seed. Germination should be prompt, and growth of the seedling vigorous. This is the seed vitality, or germinating power, and may be represented by *germination rate.* In seeds of low viability, a low germination percentage and a low rate of germination are often associated.

A reduction in seed viability and seed vitality may result from incomplete seed development on the plant, injuries during harvest, improper processing and storage, or aging. With long storage, reduced viability is usually preceded by a period of declining vitality. Seeds with a low germination rate or embryos that produce weak and abnormal seedlings are less able to withstand unfavourable environmental conditions in the seed bed than more vigorous seedlings. Weak and injured seedlings are likely to succumb to attacks by disease organisms, or they may lack the strength to emerge if planted too deeply or if adverse seed bed conditions exist. Consequently field survival is apt to be less than a predetermined germination percentage would indicate.

Measurement of Viability

If one measures the time sequence of germination of a given lot of seeds or the emergence of seedlings from a seed bed, one usually finds a pattern that is the germination curve. There is an

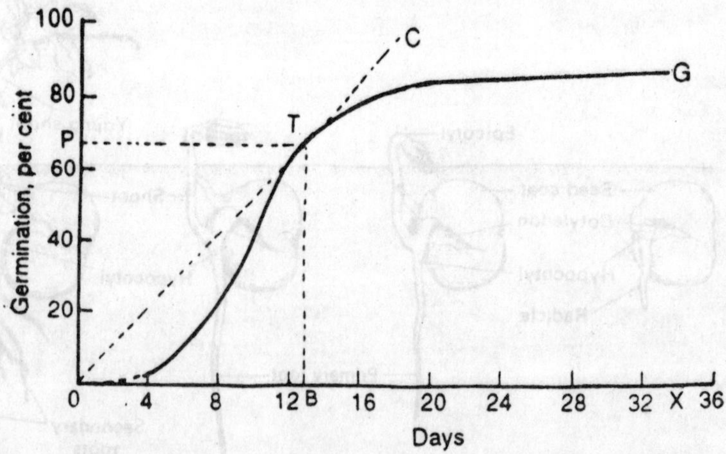

Fig. 5.6. Typical germination curve of a seed sample. After an initial delay, the number of seeds germinating increases, then decrease. Such a curve can be used to measure germination.

initial delay in the start of germination, then a rapid increase in the number of seeds that germinate, followed by a decrease in the rate of appearance. When viability is less than 100 percent, the exact end point may be difficult to ascertain. Measurement of germination involves two factors–the *germination percentage* and the *germination rate.* Germination percentage should involve a time element, indicating the number of seedlings produced within a specified length of time. Germination rate can be measured by several methods. One determines the number of days required to produce a given germination percentage. A better method, that has been used for many years, calculates the average number of days required for radicle or plumule emergence as follows:

$$\text{Mean days} = \frac{N_1T_1 + N_2T_2 + \ldots N_xT_x}{\text{total number of seeds ger min ation}}$$

N values are the numbers of seeds germinating within consecutive intervals of time; T values indicate the time between the beginning of the test and the end of the particular interval of measurement. Kotowski has used the reciprocal of this formula multiplied by 100 to determine a *coefficient of velocity.*

Czabator has suggested another measurement that includes both rate and percentage- *germination value* (GV). To be obtained by periodic counts of radicle or plumule emergence. The important values on the curve are T–the point at which the germination rate begins to slow down–and G–the final germination percentage. These points divide the curve into two parts–a rapid phase and a slow phase. Peak value (PV) is the germination percentage at T divided by the days to reach that point. Mean daily germination (MDG) is the final germination percentage divided by number of days in the test. Then GV = PV × MDG.

Regulation of Germination

A seed enclosed within the fruit normally will not germinate while still attached to the plant on which it is produced. Certain highly specialized plants–for example, the mangrove, a swamp-growing tree–are exceptions in that they produce *viviparous* embryos that germinate on the tree. These grow into massive seedlings, with a foot-long, javelin-shaped root; such seedlings eventually fall and become embedded in the mud below. Also, wet weather during harvest in certain seasons may result in premature sprouting in some grain crops, such as wheat.

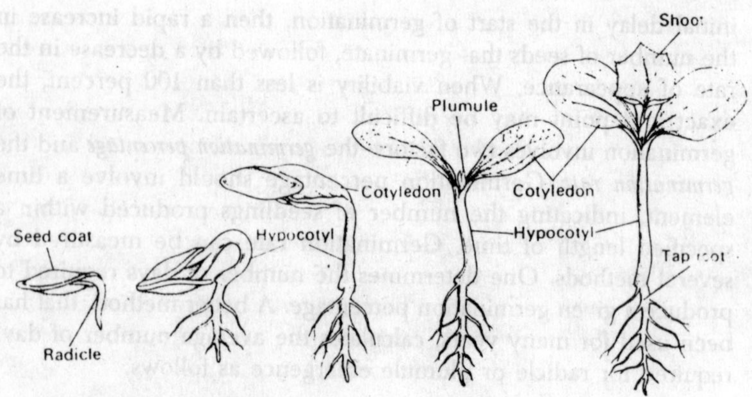

Fig. 5.7. Successive stages of epigeal germination of dicotyledonous exalbuminous seed of gourd.

The tendency toward *vivipary* is inherited as a defective characteristic and is automatically selected against. Ripening of seeds involves the development of internal mechanisms that function to regulate the time of germination so it will coincide with periods having environmental conditions most likely to favour survival of the seedlings. During maturation most seeds lose moisture to a level below that required for germination. The seed can remain in this dry state for long periods of time and can be handled with relative economy, ease, and safety. If the embryo is merely quiescent and no other germination block exists, the seed can germinate immediately whenever sufficient water is absorbed and suitable temperature and aeration conditions prevail. On the other hand, most freshly harvested seeds have additional germination regulating mechanisms that exist within the seed, and germination may fail to take place even under seemingly favourable environmental conditions.

Propagators of cultivated plants have long recognized these germination delaying phenomena and have learned to cope with them through the adoption of appropriate pre-germination and handling procedures arrived at through trial and error. Furthermore the domestication of many seed-propagated cultivars has undoubtedly included selection for easier propagation. Terminology based on these empirical methods has depended to a large extent upon the particular group of propagators. On the other hand, investigations of the fundamental nature of these phenomena have been carried out only in comparatively recent times, and an understanding is still far from complete. Such investigations have been extended from the cultivated plants to their relatives in nature, where such phenomena are not only much more prevalent and complex but play an important role in ecological adaptation.

Seeds may be grouped into the following broad categories on the basis of their response to particular environmental conditions and to methods of handling.

Group–I. Seeds in which the covering is hard and moisture-impermeable and in which the embryo thus cannot absorb water. The group also includes seeds with coats so hard as to resist embryo expansion.

Group–II. Seeds which have a dormant embryo that responds to pre-germination chilling.

- A. Seeds requiring a single chilling period.
- B. Seeds requiring a warm period (for root or embryo development) prior to the chilling period.
- C. Seeds requiring two consecutive chilling periods, separated by a warm period.

Group–III. Seeds which combine an impermeable seed coat with a dormant embryo.

Group–IV. Seeds that contain chemical inhibitors that can be removed by leaching.

Group–V. Seeds that are dormant when freshly harvested but become germinable upon dry storage (post-harvested dormancy).

- A. Seeds in which germination is promoted by light.
- B. Seeds in which germination is inhibited by light.
- C. Seeds in which germination is inhibited by high temperatures.

The term *dormancy* has broad applications in plant physiology to refer to lack of growth in any plant part due either to internally or externally induced factors. Seed technologists, on the other hand, have used the term in a somewhat narrower sense to refer to nongermination resulting from conditions within the seed (other than nonavailability). The Association of Official Seed Analysis, in the "Rules for Testing Seeds," further distinguishes between "hard seeds" and "dormant seeds". *Hard seeds* include those (groups I and III, above) which cannot absorb moisture because of the impermeable seed coat. *Dormant seeds* include those (Groups II, IV, and V, above) which are able to absorb moisture but fail to germinate because of restrictive influences within the seed that block some physiological reaction in the embryo and prevent the initiation of germination.

Conditions that exist within the seed at the time it matures on the plant to prevent germination have been called *primary dormancy.* The physiological changes that occur within the seed so that germination can take place are referred to in horticultural literature as *after-ripening.* Dormancy may be transitory, lasting only for a few days, disappearing with dry storage (in the case of seeds in group V, above); on the other hand, dormancy removal may require complex and time consuming treatments. Once a seed has passed through an after-ripening period it may again become dormant if the imbibed seed is subjected to particular unfavourable environmental conditions. This is referred to as *secondary dormancy.*

Germination controlling mechanisms are important in nature because they contribute to natural survival and to the dissemination of plants to different areas. Such mechanisms are particularly important in plants growing in desert or cold regions where environmental conditions may not be favourable for germination immediately following dissemination of the seed. For instance, in the temperate zone, spring-ripening seed on such trees as elm, maple, or willow germinates immediately, and the seedling establishes itself the same season.

If fall-ripening seeds of trees growing in the same region germinated immediately, the seedlings would probably be killed during the winter. Seeds of many plants of the latter type require winter chilling for germination, and consequently do not germinate until the following spring. Variation among individual seeds of a given lot within a particular species may distribute germination over a period of years. If those which germinate at a certain time are lost, other seeds remain to germinate at a later and perhaps more favourable time. With cultivated plants, prevention of premature germination on the plant or during seed harvesting is beneficial. On the other hand, expensive and time-consuming pre-germination seed treatments required to obtain good, uniform germination are undesirable.

If reliable pre-germination treatments are unknown, germination may be unsatisfactory and reduced stands may result. Consequently the development of cultivated crop plants has involved the selection of cultivars which do not have complex germination problems. Most difficulty in germination regulation in propagation is found in seeds of trees and shrubs of native plants or those recently introduced into cultivation. Particular difficulty is experienced in getting quick and reliable results in testing seed for viability, especially when this is attempted shortly after the seed is harvested.

Internal Conditions Affecting Germination

Seed Coverings

Impermeability to Water. The maturation of seeds of many species results not only in reduced water content but also in deposition of particular substances in the seed coverings to produce impermeability to water. This is a genetic characteristic of certain plant families, but it can also be modified by particular environmental conditions and methods of handling. A hard, impervious seed coat has value in prolonging storage life, since the internal parts of such seeds, once dry, are held in sealed storage until the seed coverings become modified. The effective seed covering may be only the hardened outer seed coat, or it may include various parts of the pericarp which either are attached to the seed or surround it.

A number of plant families contain species whose seeds have hard coats; these include the Leguminosae, Malvaceae, Cannaceae, Geraniaceae, Chenopodiaceae, Convolvulaceae, and Solanaceae. The plant group most commonly associated with harseededness is the legumes, in which are found clover, alfalfa, and similar agricultural plants, as well as locust (*Robinia*), *Acacia*, and other woody legumes.

Seed coats can sometimes be modified during seed development or by the treatment the seed receives at harvest or during storage. For instance, white clover seeds are hard when ripened in hot, dry weather but are soft when ripened during rainy weather. If seeds are harvested before they are completely ripe and before the covering has become hard, and then kept from drying out, hareededness may never develop. Further changes in the degree of hardness can take place during storage. For instance, in studies with *Symphoricarpos*, as the length of time after harvest increased, the length of the sulfuric acid treatment required to soften the coats also increased.

In some cases, seeds which normally do not have hard seed coats, such as beans, develop hareededness with dry storage and when planted, fail to germinate unless they are pretreated. Seeds with hard coverings may be germinated by any method which artificially breaks or modifies the seed coat, provided that another type of dormancy is not also present. In nature, softening of the seed coat comes about through agencies of the environment: mechanical abrasion, alternate freezing and thawing, attack by microorganisms in the soil, or passage through the digestive tract of birds or other animals. Fungus activity takes place at moderate temperatures–50°F (10°C) or higher–but does not occur to any appreciable extent at lower temperatures. For effective seed coat decomposition, the seeds must be held moist at warm temperatures in the soil. Addition of nitrates to the medium has increased seed softening, presumably because it stimulates fungus activity.

Mechanical Resistance to Embryo Expansion. In most seeds, once water is absorbed, the expanding pressure of the embryo during germination is sufficient to rupture the seed coats. However, the seed coverings can be so hard as to offer mechanical resistance to embryo expansion. In the olive, for instance, the seed is surrounded by a thick, bony, indehiscent endocarp which is both mechanically hard and water-impermeable. The hard seed covering of *Symphoricarpos* is indehiscent, but water can be absorbed through the micropyle. Other seed structures, such as the "pits" of stone fruits or the shells of walnuts and other nuts, are extremely hard and may produce temporary resistance to germination. These structures are dehiscent, however, and water can be absorbed through the layer separating the two halves. Seeds of certain weeds, e.g. *Amaranthus retroflexus* and *Alisma plantago*, also have hard coats of this type.

The force required to overcome resistance of freshly harvested seeds of black walnut (*Juglans nigra*) and hickory (*Carya* sp.) was shown to be considerably greater than that of the calculated breaking force of the embryo during germination. However, shell resistance decreased markedly during moist, warm storage, provided that the medium was unsterilized. Some structures within the seed, such as the endosperm, are believed to mechanically restrict embryo expansion in some kinds of dormant seeds. Germination occurs when the expanding force of the embryo increases or when the restrictive force of the covering decreases, or both.

Restriction of Gaseous Exchange. Physical restriction of the movement of oxygen to the imbibed embryo by the inner membranous seed coat or an enclosing nucellus or endosperm layer has frequently been cited as a factor in seed dormancy. Evidence for this is that freshly harvested dormant seed, which are light-sensitive or temperature-sensitive, can be made to germinate promptly by excising the embryo or by physically or chemically altering the seed coat. Increasing the oxygen content of the surrounding atmosphere will stimulate germination of the intact seed in many cases. The benefits of after-ripening in dry storage have been attributed to increase in permeability. The regulatory function of these seed coverings in dormancy may be real, but the mechanism does not seem to be simple.

Other factors, in addition to physical restriction, may be involved. For instance, differences in oxygen requirements of different parts of the seed or at different stages of after ripening may be involved. Cocklebur (*Xanthium*) seed illustrates this phenomenon. Two seeds are present in the bur, one germinating readily, the other delayed for one or two years. This difference was first attributed to a difference in inner seed coat permeability to oxygen uptake. It has since been found that the embryos contain two water-soluble chemical inhibitors that will not pass through the seed covering. However, the inhibitors can be removed by leaching the excised embryo with water or by subjecting the seed to an increased concentration of oxygen.

Dormant Embryos

Seed with a Single, Low-Temperature Requirement for Germination. Plant propagators have known since early times that moist seeds of certain fall-ripening tree and shrub species of the temperate zone must be chilled as they overwinter in the ground before they will germinate in the spring. This requirement led to the horticultural practice termed *stratification*, in which seeds are placed between layers of moist sand or soil in boxes (or in the ground) and exposed to chilling temperatures either out-of-doors or in refrigerators. A more accurate term for this procedure, used by some propagators, is *moist-chilling*.

Conditions required for after-ripening changes to take place within these dormant seeds include : (a) temperatures just above or slightly below freezing; (*b*) imbibition of moisture by the seed; (*c*) aeration; and (*d*) a certain length of time. After ripening is considered to be most effective at approximately 35°to 45°F (2° to 7°C), although changes also occur at higher and lower temperatures. De Haas and Schander have shown that the lowest temperature at

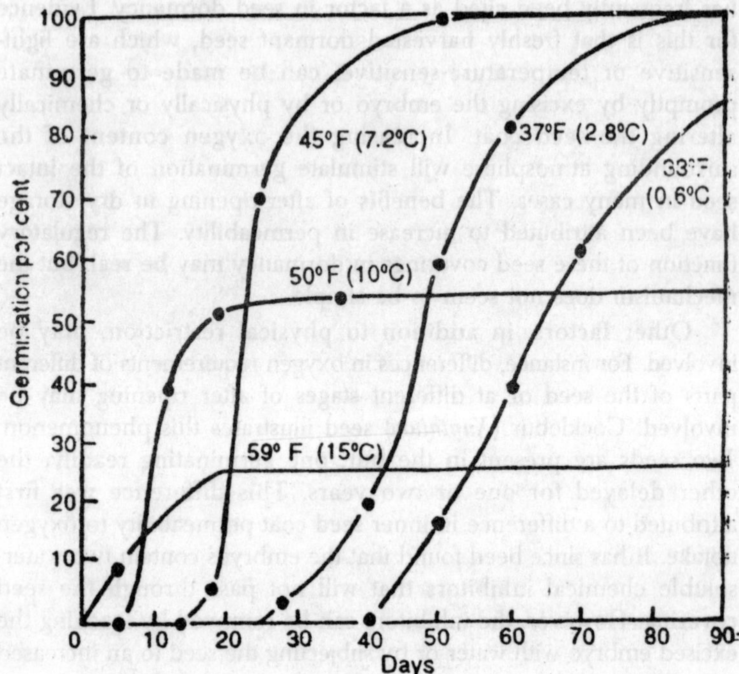

Fig. 5.8. The effect of temperature on germination of apple seeds previously stratified for 65 days at 37°F (3°C). The abscissa represents the days following stratification.

which such changes occur in apple seeds is about 23°F (–5°C). Moist, dormant seeds of many species that are exposed to a high temperature (greater than the effective chilling range) before after-ripening is complete not only fail to germinate but also develop secondary dormancy. Abbott has shown that in apple seed at 62°F (17°C) the processes leading to after-ripening and those resulting in secondary dormancy are in equilibrium. He suggested that this *compensation temperature* should be used to test germination of dormant seeds. Compensation temperatures occur in dormant seeds of other plants, but the exact temperature probably varies with different species and possibly with the stage of after-ripening.

If the seeds are placed at relatively low temperatures, the germination rate is slow but the germination percentage high. At higher temperatures germination rates are faster, but the germination percentage decreases in proportion to the increase in temperature; nongerminating seeds revert to secondary dormancy. The temperature conditions most effective for germination may be similar to those in natural environment in which soil temperatures gradually increase with the advance of spring./ Late planting or an early hot period can inhibit germination and produce secondary dormancy.

Moisture in the seed is necessary during pregermination chilling. Drying will stop after-ripening causing secondary dormancy, or it can injure the seeds. Reduced aeration will slow or stop after-ripening and has been used experimentally to induce secondary dormancy.

The time required for moist-chilling (stratification) varies among species, among clones within a species, among different seed lots of the same kind which were grown in different areas or in different years, and even among individual seeds of a single lot. Individual differences in this requirement probably account for the temperature response. It is apparent that the time requirement for after-ripening is a genetic characteristic that can be modified by environment and conditions of production. The time required to after-ripen seeds of most woody perennial species is from 1 to 3 months, although for certain species 5 to 6 months is necessary.

Removal of the seed coverings permits germination of the excised embryo, although the response of the excised embryo tends to be sluggish and the resulting seedling dwarfed and abnormal. The basis for this germination-regulating role of the seed covering is not clear. It has been explained, on the one hand, as being due to a restriction in oxygen uptake, the level of which becomes limiting to the dormant embryo at high temperatures. It has also been explained as being due to the action of germination inhibitors which are found in the seed coverings.

Flemion showed that incompletely after-ripened peach seedlings remain dwarfed for many years if held at high temperatures but will resume normal growth after being subjected to 6 weeks of chilling. Later, Pollock demonstrated that peach seedlings were dwarfed when the apical meristem of the excised embryo was exposed to germination temperatures between 73°and 81°F (23° and 27°C) but normal when exposed to higher or lower temperatures. Dwarfing can be offset by exposure of the seedlings to long photoperiods or continuous light (provided that this action is taken before the apical meristem becomes fully dormant) or by repeated application of gibberellic acid. Treatment of nonchilled seeds of some peach varieties with thiourea will produce prompt germination, but the resulting seedlings will be dwarfed.

The biological mechanism whereby embryo dormancy is overcome within the seed is not well understood. Many changes that occur within the embryo during after-ripening appear to indicate merely the increasing ability of the seeds to germinate. Such changes include increasing water-absorbing power, increasing enzyme activity, increased acidity, and gradual changes of the insoluble complex storage materials into soluble simpler substances. Pollock has measured the increase in fresh weight, number of cells, and respiratory activity in the embryo axis of cherry seeds held for 16 weeks at 50°F At this temperature, phosphorus was increasingly found in the energy-rich compounds (nucleotides) involved in the metabolic activities of the cell, whereas in seeds held at higher temperatures, the phosphorus accumulates as inorganic phosphate.

Growth inhibiting and growth-promoting substances occur within various parts of the seed. Luckwill found inhibitors in dormant apple seeds in the endosperm, integuments and embryo in the ratio of 30 : 13 : 1, respectively. The inhibitors disappeared during chilling, and growth-promoting substances appeared. Inhibiting substances have likewise been shown to be produced in the cotyledons of *Fraxinus* seeds during the period in which the dry seeds imbibed water. Lipe and Crane identified an inhibitor extracted from the integuments of peach seed as dormin (or a compound similar to it). Dormin is a dormancy-inducing substance which has been extracted from Sycamore buds and leaves; this is also known as abcisin II or as *abscisic acid.* Similarly, an inhibitor occurs in the integuments of nonchilled birch (*Betula*) seeds. Dormin and abscisin II have been shown to induce dormancy in seeds. In each of these cases the effect of this inhibitor is decreased by gibberellic acid and, to some extent, by leaching with water. Although these inhibitors tend to diminish during chilling, there is also an increase in the content of growth-promoting substances prior to and during germination. Consequently it is difficult to ascertain specific cause-and-effect relationships. Nevertheless, there is considerable belief among plant physiologists that dormancy of the kind represented in these seeds is controlled through a balance among naturally occurring inhibitors and promoters, interacting with environmental factors such as temperature and possibly oxygen supply. Gibberellin has been cited as a major part of the promoter complex, and dormin as a part of the inhibitor complex.

Seeds with a High-Temperature, Followed by a Low-Temperature, Requirement. Several cases exist in which a single low-temperature treatment is inadequate to produce germination unless the moist seed is previously subjected to warm temperatures. In nature, ripe seeds fall to the ground, are exposed to warmth and moisture in late summer and fall, and are then subjected to chilling in winter. With this sequence, germination may be delayed until the second spring. In one group of plants–wild ginger (*Asarum canadense*), several lily species (*Lilium auratum, canadense, japonicum, rubellum, superbum,* and *szovitsianum*), herbaceous and tree peony (*Paeonia*), and a number of *Viburnum* species–only the radicle grows at high temperatures. Low temperatures must follow to after-ripen the epicotyl and cause the shoot to grow. Two to 6 months (in some *Viburnums,* 17 months) at high temperatures are required for root growth before a chilling period–of 1 to 4 months–is provided.

Another group of plants with similar requirements include those in which the embryo at the time of fruit ripening is incompletely developed; it increases in size during the warm period, then after-ripens during the chilling period. Ash (*Fraxinus excelsior*) seeds have this requirement. Holly (*Ilex* sp.) seeds consist mostly of endosperm–the embryo being small and hardly differentiated–and require a warm period for the embryo to develop. In both ash and holly, however, the problem is complicated by seed coat dormancy conditions as well.

Seeds with a Low-Temperature, Followed by a High-Temperature, Followed by a Low-Temperature, Requirement. Still more complex requirements occur in a number of other plants, mostly wild species of the temperature zone–blue cohosh (*Caulophyllum*), lily-of-the-valley (*Convallaria*), Solomon's seal (*Polygonatum communtatum*), bloodroot (*Sanguinaria canadensis*), false Solomon's seal (*Smilacina racemosa*), and *Trillium erectum* and *T. grandiflorum.* Their requirements are as follows : a cold period of 1 to 6 months to after-ripen the radicle, a warm period of 1 to 6 months for the radicle to grow, and then a second cold period of 1 to 6 months to after-ripen the epicotyl.

Combinations of Two or more Types of Seed Dormancy

Seed germination in a number of species is complicated by the presence of more than one type of dormancy the usual combination being that of seed coat dormancy combined with a dormant embryo. Such seeds have been recognized by nurserymen as being among the most difficult to handle because of the long period of time required to produce germination. This group has been referred to as "two-year" seeds, because seeds planted in the spring remain dormant and will not germinate until the next spring. If planted in the fall, they will not emerge until the second spring following planting.

Treatment of seeds with double dormancy involves the combination of procedures which overcome each type of dormancy present. In nature, decomposition of the seed coverings takes place by various agencies of the environment, as has been previously discussed. Warm stratification for several months, during which time micro-organisms act upon the seed covering, followed by cold stratification, is an effective method of handling. This explains the fact that seeds planted in spring or summer fail to germinate until the following spring. The period of handling can be shortened by treating the seed coats artificially and then stratifying. In some cases giving the seeds a short, warm stratification period between the time of seed coat treatment and cold stratification will assist considerably in the elimination of dormancy. If seeds of some woody plant species are dried following harvest, a hard seed coat develops, and germination does not take place until the second spring following harvest. If, on the other hand, the seeds are fall planted without drying, germination takes place the following spring.

Chemical Inhibitors

Substances which inhibit germination can be extracted from a variety of plant parts, including seeds, fruits, leaf sap, bulbs, and roots. Some of these naturally occurring germination inhibitors have been identified as specific chemical substances : ammonia (from beet seed), hydrogen cyanide (from amygdalin), ethylene (from ripe fruits), essential oils, unsaturated organic acids, alkaloids, (nicotine, cocaine, caffein), unsaturated lactones–coumarin and others. The existence of such substances in plants does not necessarily mean they have an essential role in controlling seed germination. Nevertheless, there is increasing emphasis on the biological role that naturally occurring inhibitors play.

During fruit and seed development many different chemical materials accumulate in the fruit, seed coverings and embryo. It is not surprising then that most fleshy fruits, or juices from them, strongly inhibit germination. This occurs, for instance, in citrus, cucurbits, stone fruits, apples, pears, grapes and tomatoes. Likewise many dry fruits and fruit coverings, such as the hulls of guayule,

Pennisetum cilirare, and wheat (*Triticum*), and the capsules of mustard (*Brassica*), and others produce inhibition. Undoubtedly these substances play an important biological role in preventing premature germination while the seeds are present on the parent plant. Where the fruit covering remains with the seed, the inhibiting effect can persist for a considerable period even after the seed is separated from the plant.

Inhibitors in seeds appear to play a direct germination controlling role in some plants. For instance, seed inhibitors are important in the ecology of certain desert plants. Such substances have been referred to as chemical rain gauges in that a heavy rain (which would insure survival of the seedling) is necessary to leach out enough of the inhibitor to allow germination to take place. A light soaking is insufficient.

Inhibitors appear to the involved in germination control of many freshly harvested seeds, although their exact role is difficult to assess. Part of the evidence is that removal of the seed coat makes germination possible. Inhibitors have been repeatedly demonstrated to occur in various seeds when freshly harvested. Such materials either disappear in dry storage or leach out at germination. Dormancy has been induced experimentally in lettuce seed by treating it with an inhibitor–coumarin–and subsequently overcome by the same treatments required for naturally dormant seed. A probable role of inhibitors was described in the discussions or gas-restricting seed coats and dormant embryos.

Environmental Conditions Affecting Germination

Water

Inhibition of water by the seed is the first stop in the germination process. The two most important factors which affect water uptake by seeds are : (a) the nature of the seed and its coverings and (b) the amount of available water in the surrounding medium. Seeds have great absorbing power, owing to their colloidal nature. In storage, seeds can absorb water from the surrounding air. Different kinds of seeds vary greatly in the amount and rate of water absorbed, either in storage or during germination. The rate of water uptake is also influenced by temperature, a higher temperature favouring an increased rate. The seed covering also plays an important role in water uptake. In some seeds, it is so impermeable to water that germination will not occur until the seed covering has been altered in some manner.

The moisture supplied to the germinating seed may affect both the germination percentage and the germination rate. Germination percentage tends to be equal over most of the range of available soil moisture from field capacity (FC) to permanent wilting percentage (PWP). Differences between species become evident as the moisture content of the soil approaches dryness (PWP). Some seeds germinate only above PWP; others can germinate below PWP. For instance, vegetable plants can be grouped according to their moisture requirements for germination as follows :

Group–1. Seeds that will germinate in soils with moisture from permanent wilting percentage (or a little above) to moisture content above field capacity.

cabbage	sweet corn	muskmelon	pepper
turnip	squash	cucumber	onion
radish	watermelon	tomato	carrot

Group–2. Seeds that will germinate in soils from intermediate moisture content to above field capacity.

snap bean	pea	endive	beet
lima bean	lettuce		

Group–3. Seeds that will germinate only in soils near field capacity.

celery

Group–4. Seeds that germinate well at lower moisture contents but show reduced germination near field capacity.

spinach New Zealand spinach

Rate of emergence from a seed bed is influenced particularly by the available moisture supply. From a point approximately halfway through the range from FC to PWP there is a decline in rate. Excess soluble salts in the germination medium may inhibit germination and reduce seedling stands. Such excess soluble salts may originate from the soil and other materials used in the germination medium, the irrigation water, or excessive fertilization. Since the effects of salinity become more acute when the moisture supply is low and the concentration of salts thereby increased, it is particularly important to maintain a salinity exists.

Surface evaporation from subirrigated beds can result in the accumulation of salts at the soil surface even under conditions in which salinity would not be expected. Planting seeds several inches below the crown of a sloping seed-bed can minimize this hazard. Maintaining an adequate continuous moisture supply can be difficult, because germination takes place in the upper surface of the

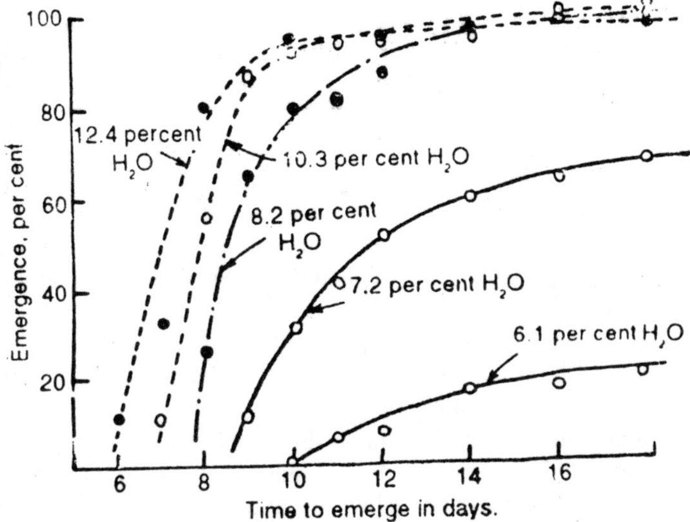

Fig. 5.9. The effect of different amounts of available soil moisture on the germination (emergence) of 'Sweet Spanish' onion seed in Pachappa fine sandy loam.

germination medium, which is subject to fluctuations in temperature and in moisture supply. The problem is greater with the necessarily shallow planting of small seeds or where the rate of germination is slow. Careful watering, deeper planting, and the application of a mulch are ways to maintain a more uniform water supply. On the other hand, excess watering accompanied by poor drainage can be deleterious, because it reduces aeration in the germination medium and favours "damping-off". Soaking seeds before planting is sometimes utilized to initiate the germination process and to shorten the time required for seedlings to emerge from the soil.

Such a treatment may be advantageous with seeds normally slow to germinate, with seeds which are hard and dry, or when certain dormancy conditions exist. However, if the seeds ordinarily germinate without difficulty, there is little need for soaking. Seeds which have imbibed water are easily injured and more difficult to plant. Prolonged soaking can result in injury to the seed, and reduce germination. The harmful effect has been attributed principally to the presence of microorganisms and to poor aeration, although there seem to be other effects as well that are not understood. If soaking is to be prolonged, the water should be changed at least every 24 hours.

Temperature

A favourable temperature is a second requirement for germination. Plants in general can be classified into the following groups according to temperature requirements : (a) those whose seeds germinate only at relatively low temperatures; (b) those which germinate only at relatively high temperatures; and (c) those which are able to germinate over a range of temperatures from cool to warm. The temperature requirement is an important factor in the adaptation of a particular species to a particular environment. Seeds of alpine plants would be expected to germinate well at low temperatures. Indeed, studies have shown that some alpines, for instance *Camassia leichtlini* and *Lewisia rediviva*, germinate only below 41°F and 50°F, respectively. On the other hand, seeds of most tropical plants require high temperatures.

Temperature largely determines the time of year at which seeds will germinate out-of-doors. As a result, seedlings of many self-sown plants, particularly annual weeds, appear only at certain times of the year, depending upon whether they are warm-season or cool-season plants.

The temperature requirements for germination of seeds are generally considered in relation to three points: minimum, maximum, and optimum. The designation of these temperatures for any one species is somewhat difficult, because temperature affects both the germination percentage as well as the germination rate. Furthermore, temperature requirements of dormant seeds may change with the age of the seed and the progress of after-ripening. Germination percentage can be rather constant within the range of temperatures at which germination takes place. The rate of germination, on the other hand, is affected more; an increase in temperature invariably increases the germination rate. Above the optimum, rates decrease as injury occurs.

Minimum temperatures are those below which germination will not occur. For "cool-season" plants, this temperature is approximately 40°F (4.5°C) or less, although for some, such as endive, lettuce,onion, parsnip, and spinach, it approaches freezing. Germination rates are invariably very low at these low temperatures. Minimum temperatures for "warm season" plants are in the range of 50° to 60°F (10° to 15°C). Below these temperatures, either the seeds fail to germinate, or "chilling injury" can result in seedling abnormalities. In lima bean, Pollock has shown that there is a temperature-sensitive period during the initial imbibition in which injury to the embryo axis occurs if seeds are exposed to temperatures below 60°F (15°C).

The maximum temperatures are the highest at which germination will occur. The upper limits of soil temperature for the survival of most vegetable seeds is between 86°F (30°C) and 104°F (40°C). Seeds placed at 113°F (45°C) were killed within 24 hours. During hot weather, direct sunlight striking the soil surface may raise the temperature to a level injurious to plant tissue. Also, moisture is rapidly lost. This so-called "heat injury" affecting young seedlings resembles the "damping-off" caused by pathogenic organisms. For may plants shading is necessary.

Short exposures can be made to temperatures much higher than those which would be used for germination. In one study, exposure of seeds of a number of different species to 120°F (49°C) for 1 hour did not decrease the germination percentage even though the seeds had imbibed water. At higher temperatures, the amount of injury increased but depended upon the kind of seed, the duration of exposure, and the moisture content of seed. Dry seeds could withstand brief exposure to temperatures of 212°F (100°C), but above 250°(121°C) all were killed.

Such heat treatments are important in freeing planting material of harmful organisms. The temperature should be high enough to kill the organisms but not so high as to impair germination. Heat tolerance is greater with seeds of low moisture content and with those that are fresh, uninjured, and of high viability.

Seeds of some species–for instance, lettuce, celery, endive, delphinium, and numerous flower species–remain dormant at temperatures over 75°F (24°C). This temperature sensitivity is associated with freshly harvested seeds and tends to disappear with after ripening in dry storage. It has been called *thermodormancy.* Germination can be induced by subjecting the seeds to lower temperatures or to dry storage. In laboratory testing, chilling moist seeds at 41° to 50°F (5°to 10°C) for 5 days will induce prompt germination when the seeds are subsequently shifted to higher temperatures. Seed germination of some woody species is also inhibited by high temperatures.

Optimum temperatures are those most favourable for germination. This should be the range wherein the highest percentage of seedlings will be produced at the highest rate of germination. On this basis, the optimum temperature for seeds of most plants is between 80°and 95°F (26.5° and 35°C).

Germination is often much better if the seeds are subjected to daily alternating temperatures rather than a constant temperature. Commonly used alternations are 59°F (15°C) or 68°F (20°C) for 18 hours and 86°F (30°C) for 6 hours. The effective temperature alternation has been shown to change as the seed after-ripens in dry storage, and the effect may disappear entirely.

The above discussion refers to the effect of temperature on the initiation of germination itself. In addition, temperature affects other influencing factors in the seed-bed, such as the activity of pathogenic organisms. Temperature likewise affects seedling growth following germination. Usually a temperature somewhat lower than that which is optimum for germination is best for growing the seedlings.

Oxygen

Respiration may be summarized as follows:

$$\begin{array}{lllllllll} \text{sugar} & + & \text{oxygen} & \rightarrow & \text{carbon dioxide} & + & \text{water} & + & \text{energy} \\ C_6H_{12}O_6 & & 6O_2 & & 6CO_2 & & 6H_2O & & 673\text{kcal} \end{array}$$

This process takes place in seeds as long as the seed is alive. In a non germinating dry seed the respiration rate is low, and little oxygen is utilized. During germination the respiration rate increases, oxygen uptake increases, and carbon dioxide is given off at an increased rate. The biochemical mechanisms that control seed germination are not well known, but oxygen is utilized in separate metabolic processes that may change with the progress of germination.

Oxygen is involved in some way with the initial triggering reactions of germination, and aerobic conditions are necessary for after-ripening reactions. Consequently limitation in the oxygen supply at the initial stages can inhibit germination. In seeds, fats are transformed into sugars, in which form they are translocated. Since a fatty acid molecule contains less oxygen than a sugar molecule, additional oxygen must be supplied. Energy to drive the cellular mechanisms and to transform specific materials into growth processes is provided through the oxidation of the reserve substances in the seed. Metabolic pathways in seeds appear to be essentially the same as in other plant tissues. Part of the energy is evolved as heat and can be measured by appropriate instruments.

Reduced aeration can decrease germination. One of the effects of a hard crust in a seed-bed is to limit oxygen diffusion and thus inhibit seedling emergence. Limitation in oxygen supply in poorly drained seed-beds or after heavy rains or excessive irrigation can result in the pore spaces of the soil being filled with water.

Seeds of land plants vary considerably in their ability to germinate under water, but those of the following species germinate well: timothy, *Agrostis nebulosa*, *Axonopus compressus*, Canadian bluegrass, Bermuda grass, lettuce, wormwood, carrot, celery, portulaca, iceplant, snapdragon, *Melissa officinalis*, bittersweet, and petunia. Seeds of water plants and rice normally germinate under water. The ability of rice to germinate at low oxygen presseures has been attributed to the presence of an anaerobic, energy-liberating system within the seeds. Cattails (*Typha latifolia*) give poor or no germination in air but germinate promptly in water or when the oxygen supply is reduced.

LIGHT

Light can play an important role in seed propagation both because of its effect upon the initiation of germination and because of its controlling influence upon seedling growth.

Light and Germination

The fact that visible light can stimulate or inhibit germination of seeds of some plants has been recognized since the middle of the nineteenth century. Plants whose seeds have an absolute requirement for light, with viability being lost in a few weeks without it, include mistletoe (*Viscum album*), strangling fig (*Ficus aurea*), and *Areuthobium oxycedri*, all epiphytic plants. Seed germination of another group, including celery, lettuce, tobacco, most grasses, and many conifers is favoured by light. Seeds of another group, e.g., some species of *Phacelia*, *Nigella*, *Allium*, *Amaranthus*, and *Phlox*, are inhibited by light.

Germination control is exerted through a photochemically reversible reaction involving the response of a pigment, known as *phytochrome*, to light of particular wavelengths :

$$\text{phytochrome}_{\text{red}} \xrightleftharpoons[\text{in far-red light}]{\text{in red light}} \text{phytochrome}_{\text{far-red}}$$

(inhibits germination)	(instantaneous reaction) in darkness (slow reaction)	(promotes germination)

Exposure of the imbibed seed to red light (6400Å to 6700Å) causes the phytochrome, in it to change to photochrome$_{fr}$ which is linked in some way to reactions inducing germination. Exposure to far-red light (7200Å to 7500Å) produces an instantaneous change to the alternate form, phytochromer, which inhibits germination. In the dark, in the presence of oxygen, and at favourably low temperatures, the latter change takes place slowly. In plants under natural light the red wavelengths dominate the far-red so that the phytochrome is converted to the active $P_{far\text{-}red}$ form.

Light-inhibited seeds are less common, but evidently their response is to the same phytochrome system. Blue light may also be involved, but the reactions are not known. Light control appears to involve the seed coats or the endosperm layer, and the light requirements disappear if the embryos are excised.

Knowledge of light effects has been obtained by experiments with particularly sensitive seeds under controlled conditions. The role of light as a germination-controlling mechanism in propagation and in nature is not always clear, since response varies with the kind of seed, the conditions of seed production, and accompanying environmental conditions. In general, a light requirement is found in freshly harvested seeds and disappears with dry storage; light stimulation will not occur if the temperature is too high. The light requirement can be partially replaced by alternating temperatures, by potassium nitrate, or by kinetin, as well as by gibberellic acid and thiourea.

In the soil, light penetration depends upon the lenght of the rays. Red light penetrates to a depth of about an inch in sandy soil. Most light-favoured seeds are small. Seed germination occurs with those situated close enough to the surface for seedlings to emerge but deep enough so that water supply will not be limiting. At greater depths there is complete darkness, or only far-red rays penetrate; in this case the seeds remain dormant until the soil is disturbed.

Photoperiod can affect germination of seeds of some species. Eastern hemlock (*Tsuga canadensis*) and birch (*Betula*) for instance, are "long-day" seeds, but germination is also promoted by low temperatures. Photoperiodic control can partially replace a seed-chilling requirement and vice versa. These light-temperature requirements for seed germination evidently are involved in the ecological adaptation of the species. Seeds of some other species, such as *Nemophila*, *Nigella*, *Veronica persica*, and *Eschscholtzia californica* are inhibited by long days.

Light and Seedling Growth

Light affects the growth processes in the seedling and has an important role in its emergence through the soil. In the darkness of the seed-bed, the hypocotyl or epicotyl is elongated and etiolated, and the leaves are unexpanded. There is a hook at the end of the plumule which pushes through the ground. When the seedling shoot emerges into the light, shoot elongation decreases, the plumular hook straightens out, leaves expand, and normal growth takes place.

In the early germination stages, the seedling utilizes the reserve supply of the seed. Later growth depends upon the production of carbohydrates and other materials resulting from photosynthesis in the leaves. Light of a relatively high intensity is necessary to produce sturdry, vigorous plants. An extremely high light intensity, on the other hand, can result in high temperatures which may produce heat injury to the seedling. Consequently excessively high as well as excessively low light intensities need to be avoided. Partial shading is necessary for many plants during their early seedling growth.

Supplementary artificial light to increase the length of the photoperiod is sometimes provided in growing seedlings, particularly during winter when the day length is short and cloudy weather often prevails. Artificial light has been used as the sole illumination in indoor propagating structures for germinating seeds. The principal problems in artificial lighting in relation to plant growth are to get a sufficiently high intensity without heating and to obtain a proper balance of light quality.

Light intensities of 700 to 1200 foot candles are satisfactory, although lower intensities can be used for some shade plants. The length of the photoperiod should be 12 to 18 hours ; below 12 hours the amount of photosynthesis becomes limiting, while photoperiods longer than 18 hours may produce poor results with some plants. Where low light intensities are used, lower temperatures should also be used. With higher light intensities, higher temperatures are permissible. A certain amount of blue light is necessary to produce stocky seedlings; red light alone produces tall, "leggy" plants. Satisfactory artificial light sources are fluorescent lamps of the daylight type, 3500°K, or 4500°K white types, or a mixture of the two, or the "cool-white" alone.

Seedling Production Indoors

The production of seedlings within some type of indoor shelter is an important method of propagating plants. Seeds are sown in special germinating media and the seedlings transferred to pots, flats, or special beds and later transplanted into their permanent location. Seeds may also be sown directly into pots or other containers. This technique is used, when outdoor growing conditions are unfavourable, to produce plants for pot plants or benching in greenhouses or to start plants for transplanting later to out-of-door locations.

It is extensively used to raise vegetables for early production, to grow plants which might not otherwise be adapted to an area having a late spring and a short growing season, and to produce bedding plants for landscaping purposes. It may also be used in the nursery production of trees and shrubs. Starting seedlings indoors for outdoor growing may enable the propagator to obtain higher germination percentages because he can control environmental conditions and reduce loss from disease, insects, and other adverse conditions. To obtain these benefits certain facilities are required, and somewhat more labour may be necessary in the handling of seedlings through to the transplant stage than would be required with direct seeding. It also requires rather close attention to details and a certain degree of skill and experience. On the whole, growing seedling indoors is somewhat more expensive than direct seeding and would be justified only with valuable plant material or expensive seeds; or when seedlings are difficult to start under more vigorous outdoor conditions, or to obtain the benefits of early season production.

Facilities

The facilities required for starting seedlings indoors depends upon the number being grown. If only small lots of seedlings are to be grown, as might be done by an amateur gardener, a small box in a sunny window would be sufficient. When large numbers of seedlings such as greenhouses, cold frames, or hotbeds are necessary. Any container that will hold the germination medium and provide drainage could be adapted for germinating seeds. Useful containers are wood, plastic, or metal flats; individual flower pots; bulb pans; or plastic freezer boxes. Seeds of plants which are difficult to transplant bare-root–for instance, cucumber or melon–could be planted individually in plant bands. Polyethylene plastic bags can also be used for germinating seeds.

SEEDLING PRODUCTION

Germinating Media

There are a number of materials and mixtures of materials which are satisfactory for seed germination. Clean sterilized containers should be used. The container is completely filled with the germination medium, which is worked carefully into the corners to eliminate air pockets. Excess soil is removed with a straight board drawn across the top of the flat. The soil is then tamped firmly with a flat block of wood to a level approximately ½ in below the rim of the container. If fine seeds are to be sown, if may be desirable to water before planting.

Sowing

The method of sowing seeds depends on the size of seed. Containers with the germination medium may be presoaked, drained, and then enclosed in polyethylene to hold moisture until planted. Depth of covering varies with the kind of seed and the germinating conditions. Very fine seed may merely be dusted onto the surface. Larger seeds may be covered to one to two times their minimum diameter, assuming that good moisture conditions prevail in the medium during the first week after planting. During the initial germination period a uniform, but not excessive moisture supply is essential in the appear layers of the medium. Containers should be sprinkled lightly after sowing and then covered. A plastic or glass cover will prevent drying out, but it should be shaded from the direct rays of the sun. Some seeds, however, must be exposed to light. Placing the planted containers under intermittent mist controlled with a time clock is an excellent method of handling.

The temperature of which the flats are placed depends upon the germination requirements of the seeds. The best temperature for seeds of most plants is around 68°F (20°C). For some, 55°F (13°C) is better, and for seeds of warm-season and tropical plants temperatures up to 86°F (30°C) are preferable. Higher temperatures (over 86°F) result is rapid moisture loss and inhibition of germination in some seeds.

Handling of the Seedlings

The primary objectives from seedling emergence until the plant is placed in its permanent location are to control damping-off and to develop stocky, vigorous plants capable of being transplanted with little check in growth. After seedling emergence, the cover over the flat is removed and the seedlings subjected to increased but not intense light. Moderate to low temperatures produce stocky growth. High temperatures lead to undesirable succulent growth, excessive moisture loss, and development of damping-off organisms. General recommendations are a 60° to 65°F. (15° to 19°C) temperature during the day and 5° to 10°F lower at night for cool season plants, and 70° to 75°F (22° to 24°C) during the day, and 10° to 15°F lower at night for warm-season plants. Seedlings subjected to adequate light produce short, stocky plants rather than spindly, elongated plants, which result if they are grown in weak light. However, full sunlight during the early stages of growth should be avoided because of injury from high temperature.

The moisture supply should be constant but not excessive. Over-watering can result in poor aeration, and the resulting high humidity contributes to damping-off. When root growth is limited in the initial germination stages, frequent light waterings are necessary to keep the upper layer of the germination medium moist. As the root system develops the frequency of watering can be reduced, but the medium should be wetted completely. Watering early in the day so that plants dry before evening is desirable to reduce fungus development. Covers of closed frames should be opened at intervals to lower the humidity as the seedling become larger.

If damping-off becomes a problem, it can be controlled to a certain degree by watering at intervals with a commercial fungicidal solution, although the success with damping-off can be controlled by this means is variable.

Transplanting Seedlings

When planted closely in the medium, the seedlings soon become overcrowded and a check in growth will occur if they are not promptly transplanted. This operation comes as soon as the first two to four true leaves have developed and the seedlings are large enough to handle. A seedling should be transplanted–with wider spacing–either into another flat or individually into pots, plant-bands, or similar containers.

A transplant flat is prepared as was done for the germination medium. Seedlings are lifted carefully and inserted 1 to 2 in apart into holes made with a wooden dibble in the transplant medium. The soil is then pressed firmly about the seedlings to eliminate any air pockets about it, and the flat is then thoroughly watered. For the first few days while the seedlings are becoming re-established, the flats should be shaded, held at low temperature, and kept from drying. When the seedlings have grown so that they are well developed and have begun to crowd, they are again transplanted either into their permanent location in the field or into pots or other containers.

Success in transplanting depends to a large extent upon the previous handling. When the permanent location is in the greenhouse, environmental conditions are controlled and are not different from those to which the seedling has been exposed during development. When movement is from a greenhouse or hotbed to the open field, the operation is somewhat more critical and requires that the plant be "hardened" prior to its shift to the open field "Hardening" involves a checking of growth resulting in the accumulation of carbohydrates which makes the plant better able to withstand adverse conditions. The process can be brought on by withholding the supply of moisture, reducing the temperature, and gradually shifting from the greenhouse or hotbed to the environment of the permanent location. A cold frame or lathhouse is useful for this purpose. Flats of seedlings can be moved into them and allowed to remain for 7 to 10 days prior to placing in the field.

Before being moved into the field, the plants should be watered thoroughly. Planting is done in the field by hand or, in some cases, by transplanting machines. During transplanting , it is desirable to retain as much soil about the roots as practical to avoid disturbing the root system. Afterward, the plants should be thoroughly watered, and if practical a temporary shade provided. For the first few days, until the plants have become established, they should be watched for wilting, and watered as necessary.

The use of "booster" or "starter" solutions containing nitrogen, phosphorus, and potassium shortly before or after transplanting is sometimes beneficial in established plants in the field. Such solution should be used with caution, since if the soil is low in moisture at transplanting time, high concentrations of fertilizer can be injurious. This effect may vary with the soil condition.

Direct Seeding

Direct sowing of seeds in permanent plant locations is a technique of seed propagation often preferred to the use of special seed-beds. In general, direct seeding is more economical, since it utilizes no special plant-growing facilities and involves no individual handling of seedlings. With many plants it is the only practical method to use. It is a standard method of growing field crop, and many commercially grown vegetables. Much forest regeneration is done through direct seeding practices. Aerial broadcasting is used for large acreages. Amateur vegetable and flower gardeners may find direct seeding preferable to other methods, since it is simple and does not make demands on time and facilities. The lower germination percentages often obtained with direct sowing can be offset by increasing the rate of sowing and thinning any excess plants after germination. This procedure is feasible where the cost of the individual seed is not a factor. Of course, if much thinning is necessary, the cost of production again increases. Direct seeding results in continuous and rapid development of the seedling, with no check in growth associated with transplanting. The total time between seeding and harvest is less than by the other methods. Some plants, such as corn, cucumbers, or beans, are difficult to transplant, and are almost always planted directly in the field.

Seed-Bed Preparation

Proper seed-bed preparation is the keynote for successful field sowing. The requirements of a good seed-bed are (a) to have an initial supply of moisture to carry the seeds through the germinating and early seedling period, (b) to have a physical condition that allows moisture to be supplied continuously to the seed, and (c) to have good aeration. A problem in seed-bed preparation is to obtain the proper balance between the extremes of low moisture-high aeration and high moisture low aeration. This can be best obtained by a medium-textured loam which can supply moisture continuously without excessive drying. A light sandy soil dries out rapidly, whereas a heavy clay soil gives poor drainage and inadequate aeration, and tends to form a hard surface crust as it dries. Soil conditioning can be provided in the filed by plowing under a cover crop or animal manure, allowing sufficient time before sowing for decomposition to take place. Incorporating peat moss, leaf mold, or other organic material may be feasible on a small scale or for high value crops. The ideal seed-bed would be one in which the soil is moist but not wet, finely pulverized to a depth of 6 to 10 inch particularly in the upper 3 or 4 inch and firmed to eliminate large air spaces which would increase evaporation and water loss.

The method of preparation depends upon the size of the operation and upon the crop being grown. For large-scale operations, implements such as plows, disks, harrows, and land-levelers are necessary. For smaller operations, the rototiller type of implement is convenient, since it loosens and pulverizes the soil in one operation. For small plots, spading or forking is the time-honored system. The soil should be spaded to a depth of 6 to 10 inch and then vigorously raked. If it is dry, irrigation several days before preparation is necessary. Working the soil when it is too wet or too dry destroys its structure and leads to difficulty in aeration and water penetration later. Determining the optimum soil condition is often difficult, but as a rule of thumb, if the soil forms a tight ball when squeezed with the hand, it is too wet. It should crumble when squeezed.

Seed-bed preparation may involve soil treatments to destroy harmful insects, disease organisms, and weed seeds. The modern development of specific chemicals and effective methods of application has made this phase of operation important, if the crop has sufficient economic value to justify the cost. This cost is offset to some extent by reduction in later costs of cultivation and weed control.

Seed Planting

The time of planting is determined by the temperature requirement for germination. Cool-season crops are planted early in the spring when temperatures are low; conversely, seeds requiring higher temperatures should not be planted until after the soil warms up. The time of planting is also determined by the date at which production from the plant is desired. For instance, planting small numbers of vegetable seeds at consecutive time intervals throughout the early spring and summer to produce a continuous supply from a home garden would be more desirable than a single planting to produce the entire crop at one peak harvest.

One of the problems in field planting is to determine the rate of sowing to obtain a desired stand of plants. The final stand of the plant is determined in order to produce the maximum yield of a high-quality, plant product. Too low a stand will reduce yields ; too thick a stand will decrease size and quality. Once the desired final stand is determined, the required rate of sowing can be calculated if the germination percentage, the purity percentage, and the number of seeds per pound (or ounce) are known:

$$\text{Pounds (ounces) of seed required per unit area} = \frac{\text{Number of plants per unit are desired}}{\text{Number of seeds per pound (ounces)} \times \text{percentage germination} \times \text{percentage purity}}$$

This figure is the minimum number of seeds to use. In practice, it is often desirable to increase the rate of sowing to offset expected losses in the seed-bed. The seedlings are thinned to a desired spacing after emergence. The depth of planting depends upon the size of the seed, the conditions of the seed- bed, and the environment at the time of planting. With too-deep planting, germination may fail because the seedling is not able to emerge before the food reserves in the seed are exhausted. In addition, poor aeration may occur at lower depths. Proper depth is also related to moisture supply. Too-shallow planting will place the seed in the top soil layers, which are subject to drying. In light sandy soils, or during periods of warm weather, deeper planting is best, whereas in heavier soils and during periods of cool, cloudy weather, shallower planting is permissible. A general rule is to plant seeds three to four times their thickness, being sure this depth is well within the moist layer of the seed-bed.

Various mulching materials have been used on seed-beds. Paper of special manufacture, and clear or black polyethylene or polyvinyl plastic film have been used. These materials are laid over the surface and either removed at the time of seedling emergence or prepared with perforations for the seedling to grow through. A later innovation is a black asphalt emulsion nontoxic to plants and remains in place until disturbed through cultivation. Other materials include perlite or vermiculite treated with polyvinyl acetate. These materials benefit seedling stands by preventing crust formation, conserving moisture, controlling weeds (in some cases), and controlling seed-bed temperatures. Temperature control depends upon the type of material; clear plastic and asphalt emulsions increase day temperatures more than black plastic but lose heat more rapidly at night.

Growing Seedlings in the Nursery

Growing trees and shrubs by seeds is an important part of nursery production. Some nurseries propagate seedlings almost entirely, dealing very little in vegetative methods. This is true in the production of trees and shrubs for reforestation, roadside planting, erosion control, and similar purposes. Nurseries producing fruit trees or ornamental shrubs and trees use seedlings extensively as rootstocks upon which to bud or graft selected clones or varieties.

The discussion which follows gives a general picture of the operations involved in the handling of seeds under nursery conditions. Details of operation vary from nursery to nursery, depending upon such factors as the kind of plant, the size of the operation, the purposes for growing, the inclinations of the operator, and the location of the nursery. Nursery management is a specialized field involving many more problems than the actual propagation of plant material.

Time of Planting

Seeds may be planted in the fall or in the spring depending upon their germination requirements and upon the management practices of the nursery. Most tree species can be placed into three general categories in regard to their seed-germiantion requirements.

Group I includes species (apple, pear, *Prunus*, *Rosa multiflora*, and yew) whose seeds require moist-chilling (stratification) and that, upon the completion of after-ripening, are sensitive to high germination temperatures. Germination temperatures of 50° to 62°F (10° to 17°C) would be optimum. High soil temperature results in damping off, seedling injury, and a tendency to revert to secondary dormancy. Such seeds could be effectively planted in the fall and sprouting would not take place until early in the spring. If they are kept in stratification storage overwinter, planting should take place as early in the spring as practical, but after frost danger is over.

Group II includes most fir, pine, and spruce species and many deciduous hardwood species whose seeds require varying amounts of moist-chilling (unless exposed to light) but that, after this is over, do not germinate in the spring until after the ground becomes warm. Optimum germination temperatures are 68°F to 86°F (20° to 30°C). Seeds of these species do not germinate in the cool fall season and not until late in the spring. Such seeds can be handled without difficulty either by fall planting or by moist winter storage and spring planting.

Group III includes those species (black locust, Colorado and Norway spruce, Chinese and Siberian elm, European larch, bristlecone pine, and Douglas fir) whose seeds are not temperature sensitive and are able to germinate over a range of 60° to 90°F (15° to 32°C). Some of these species require moist-chilling or have hard coats requiring special treatment. If fall-planted, seeds of this group will usually germinate prematurely and the seedling plants can be injured by winter cold. They are best planted in the spring, the actual time to plant not being as critical as that for the other two groups.

Other species with special requirements have been described earlier in this chapter. Planting schedules for these seeds should be geared to their particular requirements.

Fall planting is particularly useful where a large volume of seeds is to be handled and adequate cold storage facilities are not available. The seasonal labour requirements is distributed more uniformly throughout the year if some of the planting is done in the fall. As soon as the environmental conditions are favourable in the spring, germination takes place, and is not delayed by late planting, resulting in better stands and a longer growing season. On the other hand, losses sometimes result from unfavourable winter conditions, drying, disease, or rodents. Also, herbicidal weed control is necessary.

Seed-Bed Culture

Seedlings of many species are grown in special out-of-doors seed-beds during the first 1 or 2 years in the nursery. A common size is 3½ to 4 in width with a walkway between seed-beds, the length varying with the size of the operation. In some cases, sideboards are placed alongside the seed-bed after sowing to maintain the shape of the bed and to provide support for glass frames or lath shade. Seed may either be broadcast over the surface of the bed or drilled into closely spaced rows. Regulating the density of planting is one of the major problems in planting a seed-bed. For economical use of space, seeds should be planted as closely together as feasible. On the other hand, overcrowding leads to greater difficulty with damping-off and reduces vigor and size of the seedling resulting in thin, spindly trees and small root systems. Seedlings with these characteristics do not transplant well.

The optimum density depends on the species and on the purposes of propagation. In other words, if a high percentage of the seedlings are to reach a desired size for field planting or for use as rootstocks for grafting, lower densities might be desired. If the seedlings are to be transplanted into other beds for additional growth, higher densities (with smaller seedlings) might be more practical. Once the actual density is determined, the necessary rate of sowing can be calculated from certain data obtained from a germination test and from the experiences of the operator at that particular nursery. The following formula is used.

$$W = \frac{A \times S}{D \times P \times G \times L}$$

where W = weight of seeds in pounds to sow per given area.

A = area of seed-bed in square feet.

S = number of living seedlings desired per square foot.

D = average number of seeds per pound at the moisture content at which the seeds are sown.

P = average purity percentage of seeds (expressed as a decimal).

G = average effective germination percentage in the laboratory (expressed as a decimal).

L = average percentage of germinating seeds that will be living trees at the end of the season (expressed as a decimal). This is a correction factor which takes into account the expected losses which experience at that nursery indicates will occur with that species.

Seeds of a particular lot should be thoroughly mixed before planting to be sure that the density in the seed-bed will be uniform. Treatment with a fungicide for control of damping-off is desirable. Small conifer seeds may be pelleted for protection against disease, insects, birds, and rodents. Seeds are planted either by hand or by machine. Depth of planting varies with the size of the seed. In general, a depth of about two to four times the thickness of the seed is a fairly safe estimate, but this also varies with the kind of seed. During early stages of development, seedlings generally must be protected against drying, heat, or cold. In the case of tender plants, glass frames could be placed over the beds, although with most plants a lath shade is sufficient. With some plants, shade should be provided all through the first season, with others, shade is necessary only during the first part of the season. Sprinkling with water to reduce ground temperature during the hot part of the day is useful. A mulch applied to the seed-bed after planting helps to protect against drying, crusting, and cold, discourages weed-growth.

It is particularly desirable with fall-sown seeds, for seeds which are to remain for long-periods in the seed-bed before germination, or for seeds sown in cold areas. Materials which have been used include sawdust, burlap, pine litter, straw, hay, ground corncobs, wood shavings, sandpaper, canvas, shredded pine cones, sand, manure, and snow. One nurseryman reports the use of old rooting media of sand and peat as a seed-bed covering. The use of polyethylene film as a seed-bed cover has been described. In this case, the seed-bed was sown in the fall, covered lightly with sawdust, well watered, and then covered with a sheet of polythylene film, on top of which burlap was placed. The following spring the film was removed as soon as germination occurred.

During the first year in the nursery, it is important to keep the seedlings growing continuously without any check in development. A continuous moisture supply, cultivation to control weeds, and proper disease and insect control contribute to successful seedling growth. Fertilization with nitrogen is usually necessary, particularly when a mulch has been applied. Decomposition of organic material can produce nitrogen deficiency. Weed control can be facilitated by careful seed-bed preparation, cultivation, and chemical sprays. Three types of chemical controls are available.

Pre-planting fumigation with methyl bromide, chloropicrin, DD, Vapam, steam, or the like is effective and also kills disease organisms and nematodes. *Pre-emergence* weed sprays are applied to the soil before crop seeds emerge. *Post-emergence* weed sprays are applied after the crop or nursery seedlings have emerged and are growing. A wide range of selective and nonselective commercial products are available. However, such materials should be used with caution since improper use can cause injury to the young plants. Not only should directions of the manufacturer be followed, but preliminary trials should be made before largescale use.

The seedlings may remain in the seed-bed for 1 to 3 years, depending upon the kind of plant. Many plants are dug at the end of the first year and placed in *transplant beds* or "linedout" in the nursery bed for further development. Deciduous plants are planted 4 to 6 inch apart in nursery rows 18 inch to 4 ft. apart. Nursery plants in rows are often undercut at the end of one year to stimulate subsequent production of a more fibrous root system. Conifers usually go into transplant beds similar to the seed-bed but at wider spacing. Spacing may be several inches apart in rows 6 to 8 inch apart. Other cases involve field planting directly from the nursery. Seedlings produced in a nursery are often designated by numbers to indicate the length of time in a seed-bed and the length of time in a transplant bed. For instance, a designation of 1-2 means a seeding grown one year in a seed-bed and 2 years in a transplant bed or field. Similarly, a designation of 2-0 means a seedling produced in 2 years in a seed-bed and none in a transplant bed.

Direct Nursery Planting

Seeds of many deciduous trees are planted directly in nursery rows rather than in special seed-beds. This procedure is used to produce root-stocks for deciduous fruit and nut varieties. Where plants are to be budded or grafted in place, the width between rows is about 4ft and the seeds are planted 3 to 4 inch apart in the row. If the seeds show poor germinability, the seeds must be planted closer together to get the desired stand of seedlings. Large seed (walnut) are usually planted 4 to 6 inch deep, medium-sized seed (apricot, almond, peach, and pecan) about 3 inch and small (myrobalan plum)about 1½ inch. This may vary with soil type. Plants to be grown as seedlings without budding could be spaced at closer intervals and in rows closer together.

If germination is low, and a poor stand results, the surviving trees may, owing to the wide specing, grow too large to be suitable for budding. Usually, no shading is required by deciduous tree plants grown in the nursery row, although other cultural operations are similar to seed-bed culture. Plants generally remain in place in the nursery row until they are to be transplanted to their permanent location. A variation of this procedure is sometimes followed with species that germinate well but require some time to grow to budding size. Seeds are either planted close together to stand of 10 to 15 seedling per ft, or planted in a narrow band about 4 inch wide in the nursery row. The seedlings remain relatively small during the first season, after which they are lifted and lined-out in a nursery row. Cherry, apple, or pear seedlings are often produced in this manner.

6

GROWTH IN PLANTS

In the previous chapter, the structure and ecology of tropical forest communities have been considered as somewhat static structures. Of course these ecosystems are continually changing through growth and reproduction, death and replenishment, and these more dynamic aspects will be covered in the next chapter. Here the physiology of growth and development, especially of the trees, the major component of those communities, will be discussed against the background of the climatic and other factors described. The emphasis is firmly on whole-plant physiology, subjects such as metabolism, nutrition, translocation and water relations being mentioned only where they have speçial relevance to the central questions concerning the responses of particular organs. How does temperature, for instance, influence the production or growth of leaves? What are the facctros which control and integrate the allocation of carbon resources to the formation of new organs, against growth or storage in existing parts of the tree?

Such topicss are likely to be of the greatest interest both to biologists and to managers seeking to understand and use tropical forest land, for in general terms they appear to be more specific to the tropics than the underlying processes such as respiration,uptake of mineral ions or protein synthesis. Since the term growth implies the self-multiplication of living material, it includes in its widest sense all the increases and changes that occur in the life of an individual. However, a convenient distinction is usually made between quantitative increases in size or weight–growth in its narrower sense–and development, which covers qualitative changes

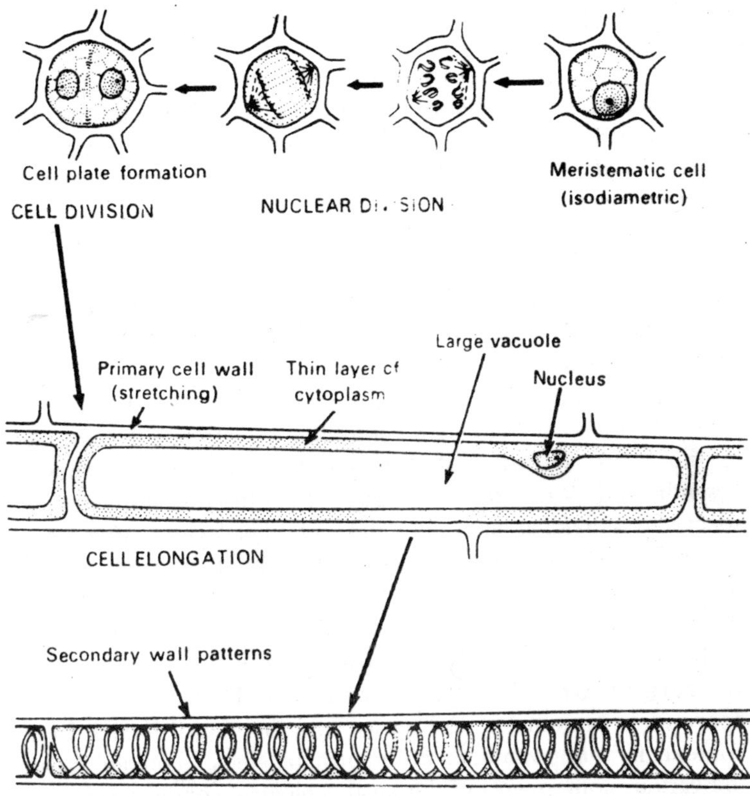

Fig. 6.1. Figure showing Cell formation, Cell elongation and Cell differentiation.

such as the intitiation of new organs, the starting and stopping of growth, and other relatively abrupt alterations to the existing physiological state. For example, measurements of the rate of expansion of a leaf blade or elongation of its petiole come into the first category; while assessments or observations on the emergence of new leaves, the duration of their period of growth and their ultimate loss from the tree involve the second. Thus the size and shape of fully expanded leaves are a function of developmental as well as growth processes.

On a given tree, they will vary because the same genetic potential has been expressed in differing environments, and also because of within plant competition and internal control systems. When comparing with a neighbouring tree, inter-plant competition, age and inherent differences may be involved as well. A logical simplification is therefore to start an experimental investigation by growing potted plants under controlled environments, keeping the conditions as nearly uniform as possible, except for the factor(s) being studied. Since genetic variability between seedlings is generally large, it is often an advantage to work with a known progeny, or better still to advantage to work with a known progeny, or better still to develop clones of well-rooted cuttings originating from seedlings. Such material can also be invaluable for research on within-tree variation, by allowing the 'temporary' influence of previous position to be separated from the more 'permanent' effects of physiological age or maturity.

Experiments of this kind cannot usually provide direct in formation on what factors are controlling growth and development in the forest setting, but this is not their primary function. Essentially, they generate knowledge about the responsiveness of a species to individual environmental and other factors. After a time, however, a sound scientific basis will be built up, from which it may be possible to predict what are the most likely ecological factors operating in the forest, and how they may interact with each other and with internal control systems.

Further experiments may then be indicated, or the most appropriate prarameters to be used in a theoretical model. In this context, it is important to distinguish between 'ultimate' factors, the various components of the local ecological conditions to which an indigenous species is normally adapted through evolution; and 'proximate' factors,some of which act as 'triggers' or signals that set in train a particular developmental change. It is often imagined that all the trees in the tropical forest are producing new leaves and stems continuously throughout the year. The fact that their shoot growth is typically periodic, with longer or shorter intervals, of activity and rest alternating with each other has been stressed since the classical studies of Schimper (1898) and Coster (1923). Nevertheless, reference to 'continuous growth in the tropics' still persists, particularly in the more journalistic or articles.

On the other hand, Kramer and Kozlowski (1960) go too far in stating that 'no matter how favourable the environment, woody plants do not grow continuously but alternate between flushes or periods of activity or dormancy'. The true position is that seedlings of many species may grow without a pause for a few months or

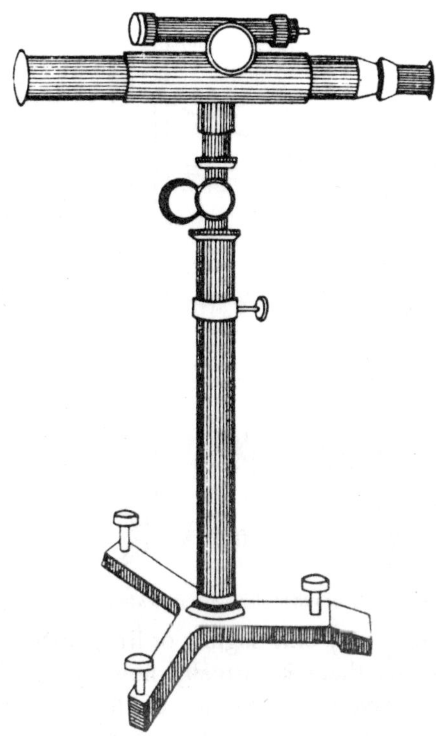

Fig. 6.2. Horizontal microscope for measurement of growth.

years, but the proportion of woody species which continue to do so afterwards appears to be less than 20 percent in Java and West Malaysia. Many of these were shrubs or small trees, and it may well be that the only groups in which continuous growth of large specimens is common in the natural habitat are the tree-ferns, conifers and palms. Once it is accepted that periodicity of shoot growth is typical of the dominant members of the forest community, the important question follows : are their periods of growth and rest appreciably synchronized or seasonal, or do they occur more or less at random?

The latter view was in fact taken by the pioneer Danish ecologist Eugen Warming (1895; 1909), who considered that 'there is no periodicity in the life of the forest as a whole'. However, subsequent work shows that it is quite usual to find seasonal peaks and troughs in flushing, leaf-fall, flowering and fruiting, not just in climates with a pronounced dry season, but in rain forests where

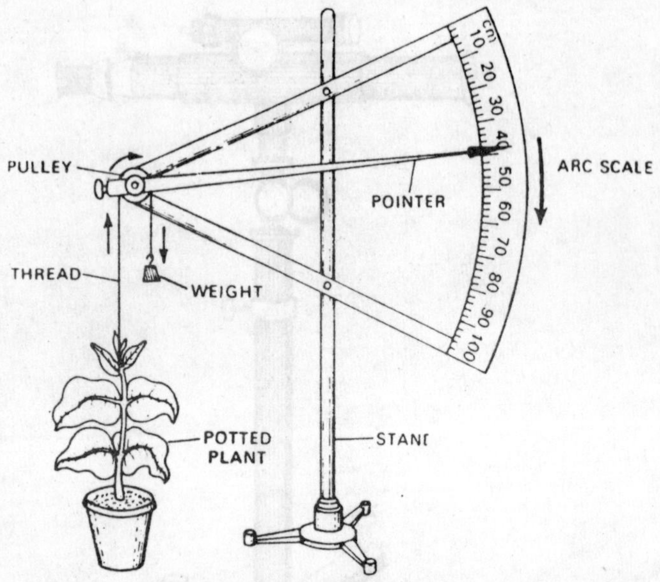

Fig. 6.3. Arc auxanometer (Arc indicator) for measurement of growth.

the conditions vary only slightly or irregularly. Indeed, Lieth (1974) considers that there is virtually nowhere on Earth that can be assumed to have no seasonality of climate. However, in order to show whether trees respond to seasons, repeated observations on tagged shoots over several years is needed, because of the presence of so many species, and considerable variation between trees of different ages, individual specimens and even parts of a single tree. Some of the seasonal, non-seasonal and irregular patterns that have been detected are described in the following sections, which deal with the growth and development of the main organs of tropical trees.

Bud-break

The first sign that shoot growth is about to start after an inactive period is often an elongation of minute leaves in terminal and some lateral positions. This may be preceded by the enlargement of the bud-scales where these are present. The subsequent, usually rapid phase of leaf expansion and internode elongation is generally referred to as flushing, and in some species is quite a striking feature of the forest which is frequently described in the literature. However, it is clear that the important physiological changes leading to renewed shoot extension must occur some time before any visible signs of outgrowth can be detected. In an early study, Simon (1914) concluded that there was probably no month without any flushing in the tropical forests of west Java, where the climate is fairly constant. The same may well be true also of some more seasonal environments, but it probably reflects the variability of the trees in the forest, rather than their uniformity. Many authors have shown that usually there is increased flushing at certain times of year, and less at others.

Some species, particularly in rather uniform climates such as that of Singapore, seem to flush regularly but to be 'out of step' with the calendar months. Others show irregular periods of bud-break, often with a great deal of variability both between and within individual specimens. Nevertheless, a general tendency towards seasonal flushing appears to be a characteristic feature of the majority of tropical forests. An example of a single-peak, annual flushing cycle, which is based on the classical studies of Coster (1928) in Java. Around twice as much flushing was found in November as in June and July, and furthermore bud-break appears to be triggered before and not at the time of the heaviest rains.

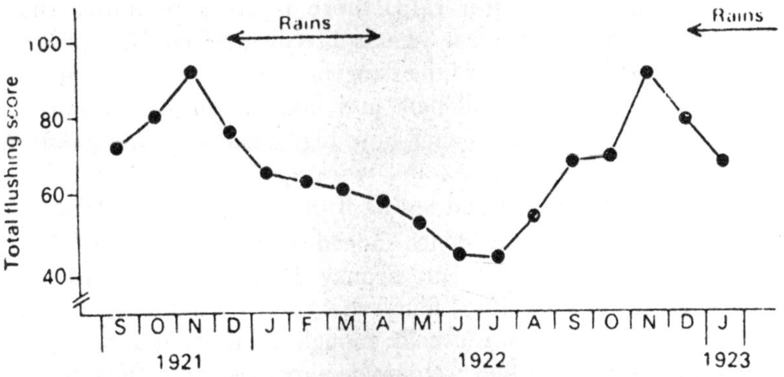

Fig. 6.4. Seasonal variation in the frequency of flushing in a tropical forest a Toeban, W. Java. 52 tree species scored on a 0-3 flushing scale.

Alvim (1964) has drawn attention to the frequency with which peak flushing coincides roughly with the equinoxes, in both Northern and Southern Hemispheres and even very close to the Equator. For example, when cocoa plants were grown under partial forest cover, they showed major flushing peaks around the time of both equinoctial periods. Other trees showing similar behaviour include *Terminalia catappa, Peltophorum peterocarpum* and *Cola acuminata.* Some trees show even more frequent recommencement of shoot growth, including cocoa when cultivated in full sunlight where the conditions are less comparable with the natural habitat of this small tree of the forest undergrowth.

Bud-break can therefore occur at any time of year, but is more likely in certain months. The pre-rains flushing already referred to is frequently though not invariably found, and flushing peaks just after the wettest season also occur. At first sight, this seems surprising, since the rains are so obviously the growing season for many crop plants, grasses and other herbaceous species. However, it is evident that for tropical forest trees as a whole this assumption may not be correct. The 'ultimate' factors which may perhaps have influenced the evolution of such seasonal replacement of the majority of the forest's assimilating surfaces will be further considered, in section 5.3, but attention will now be focused on the question of what 'proximate' factors may tigger bud-break.

It is often tacitly assumed that renewed flushing is caused by rainfall, but this clearly cannot be correct in the common situation when outgrowth starts in dry weather. Indeed, even for those tree species that flush after rain, there is often no more than curcumstantial evidence that water is directly involved. Njoku (1964), for instance, has warned of the dangers of trying to fit crop growth measurements to rainfall data and infer a causal relationship. Correlations alone, no matter how high their levels of statistical significance, only provide the working hypotheses that need experimental or theoretical testing if they are not to mislead.

In some species, evidence indeed exists that increased *water potential* (decreased water strees) may promote flushing. In teak, for example, Coster (1923) found that simply standing cut leafless shoots in pure water was usually enough to bring about bud-break in 7-10 days, whether they were inside a room or out-of-door. Similar effects were found in preliminary studies by Njoku (1963) with *Terminalia superba*, *Bosqueia angolensis* and *Millettia thonningii*. The most likely explanation for these results is that the buds were in a state of quiescence or 'post-dormancy' requiring only an increase in water content to flush, rather as a seed may lack just water for germination.

However, the same authors found that cut twigs of *Bombax malabaricum* and *Sterculia tragacantha* could only be forced in water when taken either early or late in the rest period. Thus these species may resemble many woody plants of the temperate zone in showing a period of 'true dormancy' during which the terminal buds will not flush even though conditions are favourable for short extension growth. (Lateral buds can be subject to internal control by terminals): The early and late portions of the rest period, in which outgrowth of new shoots can occur under favourable conditions, are referred to respectively as 'pre-dormancy' and 'post-dormancy'. Much more research is needs before the role played by water can be properly assessed, but two points may nevertheless be made. Firstly, many observations and experiments with well-watered potted seedlings, cuttings and mature grafts indicate that intermittent growth and rest can still occur, suggesting that other controlling factors are involved. Secondly, rainfull could have effects other than relieving water stress,such as the leaching of water-soluble inhibitors from the buds, or indirectly through the sudden temperature drops associated with tropical rain storms.

In many ways *temperature* is the most likely environmental factor to be involved in bud-break. It is perceived by all living cells, and affects virtually every chemical reaction, as well as the physical background for life processes. It is also known to have overall developmental effects: in many temperate trees, for example, chilling during the winter at temperatures just above freezing point breaks the true dormancy of their buds, while rising temperatures in the spring control their flushing in the post-dormant condition. It is not yet clear whether there are analogous systems in tropical trees, although in *Acer rubrum*, which has a very wide range from eastern Canada to South Florida, the tropical ecotype has no chilling requirement for bud burst. In cocoa (which produces repeated short cycles of shoot growth) mature cuttings flushed much more frequently in growth rooms at 30° or 31°C than at 23°C. Day temperatures appeared to be rather more important than those at night, but there was no suggestion that diurnal fluctuations as such were involved. Earlier correlations of flushing frequency with field temperature measurements had suggested that bud-break might be promoted when maximum temperatures exceeded about 28°C, and by diurnal fluctuations greater than 9°C.

A more unexpected factor requiring consideration in the tropics is *photoperiod*. It has often been assumed that because the changes in day-length occurring near the Equator are relatively small, plants might well be insensitive to them. However, it is equally possible that they might be more sensitive to small changes than species of the temperature zone, and indeed some rice varieties have been claimed to detect differences of 5 minutes. It is also known that some tropical animals are sensitive to photoperiod. Regarding woody plants, Alvim (1964) suggested that cocoa at 10-15°S showed a period without flushing because the days were shorter at that time. However, increasing day-lengths in a plantation to 15-16 h by means of bright lights did not stimulate flushing. In unpruned tea bushes growing about 27°N, on the other hand, promotion of bud-break did occur when the short natural day-lengths of November-March were extended to 13 h with weak supplementary illumination.

Experiments in controlled environments with various tropical forest tree species have shown clear sensitivity to photoperiod. Leafy seedlings of *Hildegardia barteri* remained dormant and did not flush under short-days (9 h 10 min). However, bud-break did occur when most of the leaves had become senescent and/or had been abscised, or where the plants had been leafless at the beginning of the experiments. Under long-days (17 h 10 min.), flushing took place whether leaves were present or not. Bud-break inmature cuttings of *Cedrela odorata* was also prevented by short-days unless the mature leaves had been shed. Interestingly, *Bauhinia acuminata* plants that showed earlier leaf senescence under short-days also resumed shoot elongation more quickly than the still leafy seedlings growing under long-days.

It would appear that the buds in these experiments were predormant; prevented from flushing through inhibition by the mature leaves. They could be released either by transfer to a favourable photoperiod or by loss of the leaves. In this connection, it is interesting to note that out-of-season flushing can be stimulated when *Brachystegia laurentii* is defoliated by caterpillars and that experimental defoliation induced bud-break in *Couroupita guianensis* grown in a glass-house at Heidelberg, Germany.

Besides these environmental factors and the inhibitory influence of mature leaves, it is likely that in some species endogenous rhythms are involved, though it is not necessary to invoke general hypotheses of long-term, yearly rhythmic cycles to account for the regular annual occurrence of bud-break. The internal control systems of cocoa have been particularly studied, and it has been found that periods of rapid root growth alternate with shoot flushes. When portions of the root system were removed, budbreak was delayed, suggesting that the close relationship between root and shoot activity might involve growth substances. Moreover, application of abscisic acid (ABA) to the leaves of the recent or previous flush also delayed bud-break.

When ^{14}C-ABA was applied to the leaves, it accumulated at the shoot apex, particularly during the dormant period. Conversely, applying the gibberellin GA_3 or the cytokinin zeatin promoted flushing, the radioistopes of these growth promoters accumulating at the apex most markedly during bud-burst and leaf-expansion. Further research is needed on internal factors, which might for instance control bud-break via the leaf-exchanging habit or prevent flushing in species where the vegetative buds do not grow out while a branch is flowering, such as *Ceiba pentandra* and *Hildegardia barteri.* Carbohydrate levels may be important as well as growth substances, for Fink (1982) has shown that the starch levels in stemwood of ten Venezuelan tropical tree species declined during flushing and increased again afterwards.

Rate of Stem Elongation

Two processes are involved in the production of new stems: the formation of additional nodes and the elongation of the internodes between them. Sometimes the new nodes (together with foliage leaf and bud-scale primordia) remain as a resting bud after their initiation by the shoot apical meristem, and internode elongation occurs later on. The stem units produced in a flush shoot may all have been pre-formed, or the apex may continue to form additional nodes after bud-break, so that 'fixed growth' is followed without a break by 'free-growth'. When the shoot is growing continuously for an extended period, new nodes are being formed and internodes extending without a break. This section is concerned with the many factors affecting the rate at which these processes occur, including the production of new leaves. Those influencing leaf expansion, which usually takes place at the same time, are considered. The rate of stem elongation in some tropical woody plants can be extraordinarily high.

Certain species of bamboos may even elongate pre-formed tissue at almost 1 m per day, and vines and lianes can also grow very fast. On a longer time scale, early growth rates of young trees under natural conditions in Zaire, the three climax species increasing in height at an average rate of 0.3-1.2 m per year, and the three pioneers at 1.2-3.7 m per year. It is also noticeable that the former

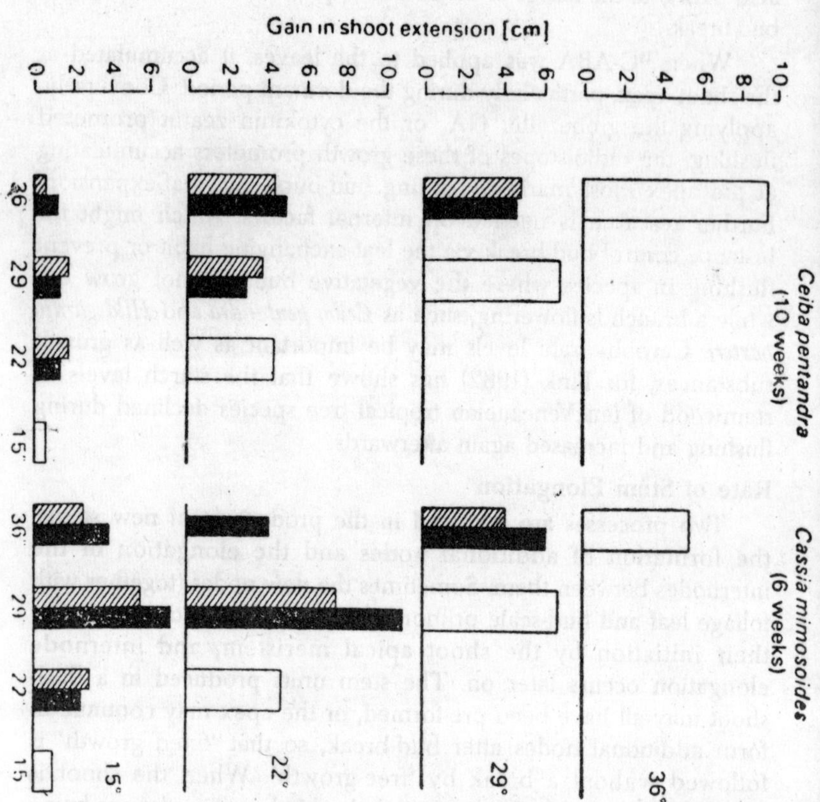

Fig. 6.5. Height growth of young trees under natural conditions in Zaire. Full line–canopy species: (1) Scorodophloeus zenkeri; (2) Oxystigma oxyphyllum; (3) Gilbertiodendron dewevrei. Dotted line–secondary species in clearings; (4) Calancoba welwitschii; (5) Terminalia superba; (6) Musanga cecropioides.

group increased their height increment as they got older, whereas rates steadily declined in the latter. In their study of over 1,500 trees during early succession of cleared forest in Ghana, Swaine and Hall (1983) found that after 5 years the majority of the late seral stage trees had grown at less than 0.5 m per year. Conversely, there were 8 species of pioneers that had exceeded an average rate of 2 m per year. Rates of height growth in the first year can be as much as 3-4 m for such heliophilous species, when thriving as seedlings or coppice shoots in large clearings or plantations. Exceptionally, 8-9 m in the first year has been recorded in *Albizia falcataria. Sesbania grandiflora, Eucalyptus deglupta* and *Trema micrantha*, the latter exceeding 30 m in 8 years. Although height growth in tropical trees can clearly be very rapid indeed, it is easy to exaggerate the general growth rates in the forest by concentrating attention on the performance of the leading shoots of a few species in particularly favourable conditions.

Most of the other shoots on a tree grow more slowly, and by the time the middle and upper tree levels are reached, stem extension, even of the most vigorous shoots, may well be reckoned only in centimetres per year. At the lower tree level, woody plants often survive with minimal increase in height for many years, while growth in small gaps is ususlly much less than in the open. The rate of stem extension can be affected by all the environmental factors discussed as well as by others such as the force of gravity and the presence of atmospheric pollutants. Not surprisingly, *water availability* is important, rates often being affected by quite small deficits. Currently expanding internodes can be inhibited, and reduced numbers of nodes may be formed during rest periods.

Cell expansion in plants clearly depends directly on turgor, while photosynthesis, metabolism and growth often slow down because water deficits stimulate an abscisic acid-induced closing of stomata that may persist for some days, even if the stress has been relieved. Daily fluctuations in growth rate of stems, as well as longer term effects, can result from alteration in plant water potential. Sixty years ago, Coster (1927) used an automatic recording technique to follow in detail the elongation of a young tropical bamboo shoot. During the night it grew at 13 mm h^{-1}, but there was a very pronounced check to extension growth during a hot day. The following day was also hot and sunny, and by 11.30 h the rate had fallen to 5 mm h^{-1}. At this point, he removed the four tall mature canes growing from the same clump, whereupon before noon the growth rate increased to 16 mm h^{-1}. Since the change was so large and so rapid, the conclusion is almost in escapable that elongation had been retarded by temporary water stress, and that it increased because the main transpiring surfaces had been removed.

In contrast, Coster found that the vine *Aristolochia gigas* grew at a rather steady rate throughout a night and a hot day, while in the liane *Congea villosa* day-time growth exceeded that made in the night. In view of such diversity of response, it would be unwise to make any general assumptions that shoot growth is typically greater at night, as is sometimes done.

Diurnal fluctuations of *air temperature* in the tropical forest can sometimes be considerable, particularly in the upper canopy. Seasonal changes are often less pronounced or minimal, and comparison with temperate zone conditions has led some to conclude that temperature might not be a particularly important factor. However, experiments with well watered potted plants in controlled environments indicate quite clearly that many tropical tree species are particuarly sensitive to small temperature differences. Thus *Ceiba pentandra* seedlings kept at 36°C grew 23 times faster than others from the same batch places at 15°C. In several experiments, increasing only the night temperature by 5-10°C promoted shoot elongation rates substantially. Cocoa appears to be a particularly sensitive species, for in one trial raising only the day temperature by 3.5°C stimulated shoot growth of 250 percent.

The full range of temperatures in which shoot growth occurs has not been thoroughly explored. Minimum temperatures may often lie in the range 12-15°C, below which many tropical plants in any case show chilling injury *Guarea trichilioides* and *Avicennia marina* appear to have a high minimum temperature of 21°C for shoot growth. Maximum temperatures for shoot elongation (and for survival) often seem to lie between 40°C and 50°C. As a general rule, the optimum temperature for shoot growth lies nearer the maximum than the minimum end of the range. A value of around 30°C was suggested for young seedlings of *Ceiba pentandra* and *Leucaena leucocephala*, which is perhaps rarely likely to be experienced in nature. Another feature of the shoot elongation of *C. pentandra* is that it is more rapid with constant rather than fluctuating temperatures.

Moreover, if the pairs of shaded and solid histograms are compared, it is clear that interchanging the day and night conditions does not make a great deal of difference. However, in the small woody coastal plant *Cassia mimosoides*, the warm day/cool night regimes were generally more favourable, and alternating temperatures of the led to more growth than constant conditions. The optimum regime appeared to be 29°C day/22°C night, close to conditions that might often be experienced in nature. Similarly, young coffee plants made most shoot growth with the combination 30°C day/23°C night.

Table 6.1 Effect of night temperature on shoot elongation of seedlings of tropical forest trees.

Species	*Night temperature (°C)*			
	Lower	*Higher*	*Rise in temperature*	*% Increase in shoot growth*
Cedrela odorota	20	30	+ 10	+ 94
Triplochiton scleroxylon	20	30	+ 10	+ 65
Ceiba pentandra	26	31	+ 5	+ 74
Ceiba pentandra	31	36	+ 5	+ 6
Bombax buonopozense	26	31	+ 5	+ 72
Bombax buonopozense	31	36	+ 5	+ 21
Gmelina arborea	26	31	+ 5	+ 140
Gmelina arborea	31	36	+ 5	- 14

Although cooler nights may well conserve assimilates, it should not be assumed that most species will necessarily be similar to the two just described. Even in *C. mimosoides*, when the day temperature is kept at 15°C or 22°C, shoot growth is promoted when nights are 7-14°C warmer than the days. *Gmelina arborea* apparently has a night temperature optimum for leaf production near to 36°C, while it is probably around 31°C for internode elongation of *Ceiba pentandra*. Some species actually have a higher night temperature optimum than that for the day, such as the tropical herb *Saintpaulia ionantha* and some provenances of the subtropical conifer *Pinus ponderosa*.

One way in which warm or hot nights might perhaps stimulate shoot growth would be if they promoted leaf production and/or expansion, and thereby increased the plant's subsequent capacity for photosynthesis. The rate of elongation by shoots may also be affected by the photoperiod under which the tree is grown. Notonly are at least 14 tree species sensitive to this factor, but the effects can be quite large. For instance, shoot growth in *Terminalia superba* seedlings was trebled by increasing the day-length from just over 9 h to the value found in June near the coast of West Africa. This species is so sensitive that statistically significant photoperiodic effect could be detected after as little as three days treatment. The effect involved both internode elongation and the number of nodes and leaves produced.

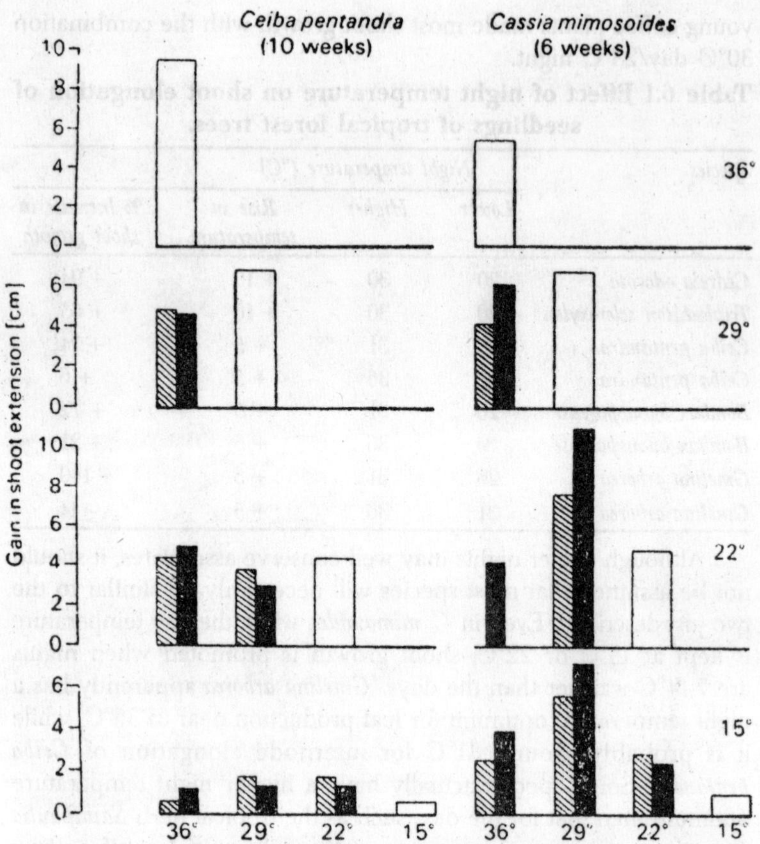

Fig. 6.6. Effect of constant and fluctuating air temperatures on the rate of shoot elongation in Ceiba pentandra and Cassia mimosoides seedlings. Open column–constant temperatures; solid column–fluctuating temperatures, with cooler night; shaded column–fluctuating temperature, with warmer night. Both day and night regimes lasted 12 h; day-length=13.2 h. Almost all Cassia plants in 22°C day/36°C night died before week 6.

The latter shoot growth parameter, and also short and total dry weights, were especially stimulated in *Chlorophora excelsa* when day-lengths were lengthened to about 17 h. All the treatments in these experiments recieved similar photon flux densities of shaded natural sunlight in glasshouses and fluorescent light in growth rooms, the day-lengths being extended as necessary by supplementary incandescent illumination too dim to influence photosynthesis appreciably. Thus it is clearly established that many tropical trees are sensitive to photoperiod, influencing either the rate of leaf production or internode extension or both. Other species in which long-days promote shoot growth include *Pinus caribaea* var. *hondurensis*, *Rauvolfia vomitoria*, *Hymenaea courbaril*, *Bauhinia acuminata*, *coffee* and *cocoa*; while extension of *Gmelina arborea* shoots appears to be unaffected by changes in day-length.

Table 6.2. Effect of day-length on shoot elongation of seedlings of tropical forest trees.

Species	*Photoperiod (h)*			
	Shorter	*Longer*	*Increase in day-length*	*% increase in shoot growth*
Cedrela odorota	11.0	14.5	+ 3½	+ 51
Theobroma cacao	12.5	14.5	+ 2	+ 31
Terminalia superba	9.2	13.2	+ 4	+ 195
Terminalia superba	13.2	17.2	+ 4	+ 59
Chlorophora excelsa	9.2	13.2	+ 4	+ 22
Chlorophora excelsa	13.2	17.2	+ 4	+ 355
Ceiba pentandra	9.2	13.2	+ 4	+ 130
Ceiba pentandra	13.2	17.2	+ 4	+ 1
Bombax buonopozense	9.2	13.2	+ 4	+ 82
Bombaxbuonoponzense	13.2	17.2	+ 4	− 7
Hildegardia barteri	9.2	17.2	+ 8	+ 200

Synergistic effects between day-length and night temperatures have been found in *Terminalia ivorensis*, and in *Ceiba pentandra* where 50 percent more leaves were formed when both promotive factors were applied, but neither had a significant effect when given alone. More complex interactiosn appeared to be involved in the control of internode length and total shoot elongation,which were both strongly influenced by day-length and night temperature. Thus the question as to whether trees under natural conditions will be influenced by photoperiod is difficult to answer. However, it is clear that shoot growth of a number of species can be substantially modified, increases of up to 75 percent per hour having been demonstrated (Table 6.2). Some evidence exists of sensitivity to alterations in the range experienced in the humid tropics, where natural day lengths vary by just over 30 minutes at a latitude of 5°C, and by more than an hour at 10°C. Besides possible photoperiodic effects, the shoot growth of forest trees may be affected by the light quality received, especially by the red/far-red ratio. The photon flux density is clearly also an important consideration, both indirectly through carbon assimilation and perhaps also directly. Effects of shading on growth are often profound, as has been shown for instance with seedlings of several species of Dipterocarpaceae.

It should be remembered that differences may not always be due to light itself, since treatments may also produce modifications of light quality, leaf temperatures and water potential. Clearly, many other factors will affect the rate of shoot growth. For instance, there was a very high correlation (r= 0.99) between the level of phosphorus fertilizers added and the height growth of *Terminalia ivorensis* seedlings. Besides external conditions, it is important to bear in mind that other processes which are proceeding simultaneously in th tree, such as root growth, may affect the shoots, as pointed out more than 25 years ago by Kursanov (1960) regarding the aerial roots of an epiphytic *Ficus* sp. Flowering or fruiting may divert carbohydrates away from shoot growth,or perhaps modify it through the mediation of growth substance.

It is also possible that sizeable differences may arise through the presence or absence of specific mycorrhizal fungi associated with the root system. Genetic differences between individuals in a population or species can be surprisingly great, as is demonstrated by the very substantial clonal variation in height growth that is typically found between clones of tropical forest trees, which in *Triplochiton scleroxylon* may be related in part ot different net assimilation rates. Higher photosynthetic rates per unit leaf area have also been reported for *Ceiba pentandra* than for *Musanga cecropiodes, Terminalia ivorensis* or *Chlorophora excelsa.*

Onset of Bud Dormancy

It is often possible to detect that a flush or a longer period of extension growth is coming to an end by observing that there are few or no expanding leaf primordia at the shoot tips. Once the terminal bud has entered some form of dormancy, stem elongation is confined to the completion of the extension of the last few internodes. The shoot apex itself is not necessarily in active, and often continues to initiate new leaves and 'stem units' during the dormant period, at least for some time. Besides foliage leaves, the apex may produce but scale primordia; and there are some species in which dormancy is signalled by the abscission of the whole of the terminal shoot tip, leaving lateral buds to continue axial growth. Such shoot tip abortion is a common feature of the shoot growth of *Cedrela odorata, Xanthophyllum curtisii* and *Citrus.* Thus the height of a tree may depend on three additional considerations, besides the factors influencing the rates of node production and internode elongation discussed in the previous section. These are the duration of the period of shoot elongation, the number of preformed nodes within resting buds and the duration of the period of bud dormancy.

Fewer ecological studies have covered the time of cessation of shoot elongation than the more obvious phenomenon of flushing. What information there is suggests that in plantation crops that grow by repeated flushes, the duration of elongation may be 6-7 weeks, for example in cocoa, and as little as 1½-2 weeks in rubber and mango. In the forest, it seems likely that periods of shoot extension of 1-2 months may become fairly genera after the seedling and sapling stages. The onset of bud dormancy will therefore occur at different times of year in trees which flush intermittently or irregularly, but in more regularly flushing species it will tend to be seasonal.

In Ghanaian forests, for example, the commonest month for the ending of shoot elongation appears to be April which is in the transitional period between the main dry and rainy seasons. Even in a species such as cocoa, which may flush several times a year it was observed several decades ago that most of the growth of mature cocoa occurs in the drier months ... while during the main wet season when conditions of water suply and humidity are most stable,little or no growth occurs.

In a tropical dry forest in Costa Rica, production of new leaves and shoots declined as the rainy reason advanced, and in a wet evergreen forest it decreased as the drier part of the year passed.

Although there are exceptions, it is remarkable that many tree species should be ceasing shoot elongation at a time when the 'growing season' is just starting. There is a parallel here, however, with the situation in the north temperate zone, where many species cease producing new leaves in May, June or July, when the environment for growth seems ideal. As was pointed out a century ago, trees often re-start shoot growth under less favourable temperature and other conditions than those obtaining when they stop elongation. However, perennial plants might well gain an important selective advantage by producing new foliage in the latter part of the dry season. Unlike annual grasses and many short-lived crops,their recently expanded leaves, at the peak of their photosynthetic capability, will then be present at the time of the greatest daily totals of photon cloudiness in the rainy season, and sometimes by dust or haze at the height of the dryest season, giving sunlight maxima around the equinoctial periods. An additional 'ultimate' factor that could be operating is that a leaf expanding in the latter part of the dry season may be subject to less overall water stress than one formed at any other time.

If the dormant period in many mature trees is as long as 10-11 months, this will curtail the yearly height increment even when the rate of elongation is rapid. In seedlings too, growth of *Triplochiton scleroxylon* and *Hildegardia barteri* fell behind that of *Musanga cecropioides* mainly because they became dormant for a while, when the *Musanga* grew continuously. This may also be one reason for the decline in height growth often found with increasing age : for example 3-year-old *Bombax buonopozense* stop height growth for only 3-4 months, whereas older trees are dormant for 9-10 months.

The physiological signals which induce a change from active shoot growth to terminal bud dormancy appear to include both external and internal factors. For instance, it has been demonstrated using controlled environments that shoot extension can be halted in a number of tropical tree species by reducing the *photoperiod.* Seedlings of *H. barteri* stopped extension growth quickly under 11½ h days, and more slowly in 12 h days. When given 12½ day-lengths, however, they grew continuously throught out the 26 week experimental period. Very rapid onset of dormancy also occurred inleafy *Cedrela odorata* plants exposed to short days, but theprocess was slower in *Terminalia superba* seedlings, about 50 percent of the treated trees stopping temporarily, while the other half continued to grow slowly. A proportion of *Ceiba pentandra* seedlings grown under short photoperiods also showed temporary dormancy. The length of the shoot extension period was decreased by about 6 weeks in *Bauhinia acuminata* when day-lengths were shortened, but a reduced daily light energy supply or temperature differences may have been involved as well as photoperiod in this experiment using a thick tarpaulin screen to apply the short days.

Cooler night *temperatures* tended to induce bud dormancy in *Gmelina arborea,* some plants stopping height growth with nights of 26°C, but none doing so at 31° or 36°C. Cool nights caninteract with short-days, as in *Bombax buonopozense.* Under natural conditions in the forest, a reasonable hypothesis might be that warmer nights and longer days might promote faster and more prolonged shoot extension, while cooler nights and shorter days could lead to reduced shoot growth and to cessation of height growth. However, it is very probable that other external factors will hasten the onset of terminal bud dormancy,for example reduced light intensity, a low red/far-red ratio, water stress and shortage of mineral nutrients.

Internal conditions within the tree also appear to be important incontrolling the onset of dormancy. Species which show repeated flushed of growth, such a cocoa, tea, mango and rubber, are good examples of this, although it should not be assumed that they will be unaffected by their external environment. One possibility is that a rapidly growing shoot might stop simply because new leaves were not being initiated quickly enough by the apes, which may happen when tea is grown in the tropics. Alternatively, each shoot flush may be confined to those leaves and stem units pre-formed in thebud. In this case, dormancy is as it was predetermined, perhaps withan 'articulate' shoot morphology reflecting the cyclic production of bud scales and foliage leaves.

A series of detailed studies on thc photosynthetic capacities of different leaves of cocoa, and the translocation of assimilates within the plants, has shown that first year plants are flushing rapidly and expanding many new leaves, they are using much more carbohydrate than they are producing. Thus, it is possible that lack of assimilates might lead to cessation of shoot elongation; for example, successive removal of expanding leaves can prolong shoot elongation periods, for instance in rubber and in *Couroupita guianensis.* In temperate species of the tropicallly originating genus *Quercus,* it is even possible to turn an intermittently flushing tree into one which grows continuously. Some instances are also known in which individual specimens of intermittent tropical species exhibit continuous growth for long periods, as invigorous leading shoots of unplucked tea. Examples such as 'lamp-brush' rubber and 'foxtail' pine may be due to genetic differences, perhaps becoming expressed when the plant is grown out of its natural environment.

It is likely that there are often negative feed-back control systems that change or maintian the phases of shoot elongation or terminal bud dormancy. The interactions between roots, leaves and buds may well involve changes in the endogenous levels and distribution of both growth promoters and inhibitors, and changes in levels of these have been detected at different times in the flushing and dormant cycles of cocoa. In this species, internal water stress was apparently not involved, but it has been implicated in studies with other tropical trees.

Growth of Leaves

Young, growing leaves quite often attract attention in tropical forests because their colour or appearance is different from that of the mature foliage. They are often a lighter shade of green,because the formation of chloroplasts and the synthesis of chlorpohyll *a* and *b* has only just started. Sometimes they are very pale– for example in the half-expanded flush shoot of cocoa grown under shade–and they can even appear white, yellowish or blue in some species. The commonest colouration is red, generally caused by anthocyanins in shade to a bright colour which masks any green present, as seen for example in *Cynometra ananta* and *Carapa procera.* In some species, such as *Amherstia nobilis,* the new flush of pale-coloured young leaves all hang downwards for some time, which may be an adaptation that reduces waters tress until the leaves are capable of active photosynthesis.

A feature of tropical trees which is fairly common is the capacity for active leaf movements. The orientation of leaves and leaflets can be altered, not just by growth curvature of the petioles, but at swollen *pulvini* or 'leaf-joints' at the base of the organ in question. The turgor changes which are responsible for the movement can be relatively rapid, so that the plant reacts quite promptly to a change in environment. The best-known example is the 'sensitive plant', *Mimosa pudica,* which is a cosmopolitan weed of open spaces in the tropics. here the reaction time to the stimulus of a leaflet being touched is of the order of a second or less, and the leaflets also close (more slowly) in darkness. Light at wavelengths of 403 nm (violet) or 726 nm (far-red), but not at 585 nm (yellow) or 656 nm (red), has been found to cause opening of leaflets, but this was only effective when given between 0600 h and 1600 h, indicating the existence of an endogenous diurnal rhythm. Applied indole acetic acid also led ot opening, even in darkness, and autoradiography showed that it could spread from the rachis throughout the pinna in 4 minutes. In *Albizia,* a red/far-red seversible system has been shown to control opening and closing of the leaflets, the turgor changes in the pulvinus apparently being triggered by a massive influx or efflux, respectively, of pottassium ions.

In many leguminous trees, the leaflets close at night, and sometimes also in response to other stimuli–for instance those of *Brachystegia spicaeformis* have been shown to close during the day when leaf surface temperatures surpass 32-34°C while dark closure was promoted and the duration of the closed phase increased in *Cassia fosciculata* when temperatures were reduced form 35° to 15°C. Movements are not confined tothe Leguminosae, being noticeable at night in *Triplochiton scleroxylon*, for example, while an unusal example is the formation of 'night buds' in *Espeletia sahuttzii* at 3600 m in the Vanexuelan Andes. The rosettee of leaves close around the apical buds at night, which has been shown by measurement with thernoccuples to protect the young leaves from freezing temperatures, and from rapid heating up just after sunrise, either of which can be lethal to them, but not to the older leaves.

The physiological significance of changing leaf orientation probably lies in the achievement of a successful balance between maximal photosynthesis and minimial water stress. It addition, as noted, it has important ecological effects on other leaves, by altering the proportion and the quality of light that reaches them at different times. Similar consideration apply to the regular occurrence of upper canopy leaves that are more or less permanently oriented at a steeply inclined angle, rather than horizontally. The leaf mosaic can also change rather abruptly under a previously leafless tree crown that suddenly produces a new flush of leaves, and there are continual, slighter modifications resulting from the expansion and loss of individual leaves,or grazing by herbivores.

Seasonal changes which affect the forest as a whole may also occur. Peaks of new leaf growth generally occur shortly after those for bud-burst, because the expansion of leaves typically occurs at the same time as internode elongation. Sometimes the two processes appear to be closely linked; thus, for example, in tea a single deciduous bud scale amongst expanding foliage leaves will have a correspondingly short internode proximal to its insertion. Often the first foliage leaves on a new shoot are smaller and may be similarlyl associated with shorter stem units. On the other hand, the first internode of a sylleptic branch is typically much longer thanteh subsequent ones, but the leaves are usually of a similar size. In *Terminalia ivorensis*, for example, experimentally reducing the are of young expanding leaves did not significantly affect the internode lenghts.

Growth rates of leaves are often rapid, especially at the time when they are about half their half size. Leaflets of *Amherstia nobilis* can increase in length by 18mm per day, and petioles by 41mm per day. Leaf growth in this rapidly flushing species is sometimes completed in less than two weeks, although it takes another fortnight for the leaves to spread out horizontally. The rachis of the large fern *Angiopteris evecta* can even extend at 90mm per day, and Coster (1927) found that leaf night when the weather was sunny, but not on a cloudy day. In the great majority of species, leaf growth is determinate, so that in some species with large,compound leaves each one may take several weeks to complete its growth. In a few instances, such as *Guarea guidonia* and *Chisocheton* spp., the leaf-tips remain meristematic, producing new leaflets continuously or periodically over a longer period of time.

Leaf growth usually involves both cell division and cell enlargement, and the final size reached, and sometimes also the leaf shape, can be affected by a range of external, internal and genetic factors. A striking example of unequal growth of adjacent leaves is provided by plants showing strong *anisophylly*, for example the tropical American *Columnea sanguinea*, the Malaysian *Anisophyllea trapezoidalis* and the West Africal *A. laurina.* The branches have two rows of large leaves below, and two of very small leaves above, while on vertically growing main stems all leaves are large. A possible hypothesis might be that the horizontal position of the branch in relation to the gravitational field imposes a *gravimorphic* effect, so that either inhibition of growth occurs inleaves inserted on the upper side, or the two types of leaf arise from primordia that contain very different numbers of cells. Less pronounced anisophylly is found in other genera, for example, *Alstonia* and *Gmelina.*

Water stress is one of the factors likely to reduce leaf growth, possibly through the production of large amounts of ABA which may close stomata. In cocoa, for instance,when the available water content of the soil had decreased by a third, leaf growth had slowed considerably. By the time the moisture supply had decreased by a further third, expansion had stopped entirely, so that leaves only reached a length of 25-60 mm, while the contrly leaves on plants in moist soil were twice as long and still growing.

Shading commonly influences leaf development, generally increasing the rates of growth and final sizes, and sometimes modifying leaf shape. In cocoa and coffee, for instance, which are small trees of the forest understory, the leaves that expand in full sunlight are often smaller and yellowish-green, while those growing in shade tend to become larger and darker green. Increasingly, it has been found that cocoa leaves in full light grow considerably larger if the lower mature leaves on the same shoot have been removed (Jan Krekule, unpublished data). The implication is that the effects of light intensity and quality may be indirect perhaps operating via water stress. Even in rigorously controlled environments, it is quite difficult to distinguish between genuine control by photon flux density and secondary effects such as water stress or leaf temperature changes, or tertiary responses such as the closure of stomates. In shading trails in the field, differing air and soil temperatures also have to be taken into account, together with variation in atmospheric saturation,vapour pressure deficits and any root competition from shade trees.

Temperature and photoperiod can affect leaf growth; thus seedlings of *Triplochiton scleroxylon* grown under 11 h days + 20°C nights produced leaves that were 35-40 percent shorter than on plants receiving either 14½ days, or 30°C nights, or both. *Ceiba pentandra* and *Gmelina arborea* also showed an interaction between these two factors, while in *Chlorophora excelsa* leaves were 50 percent longer when grown under long rather than short-days. The effects on leaf area were likely to have been even greater than is suggested by these assessment of linear dimensions. Leaves of a paler, yellowish-green colour tended to occur when *Triplochiton scleroxylon, Terminalia ivorensis* and *T. superba* plants were grown in cool day/hot night regimes, in comparison with other combinations of fluctuating or constant temperatures.

There are likely to be many other influences upon the development of a leaf, including such factors as the availability of nitrogen and mineral nutrients, and the influence of herbivore damage to it (and elsewhere in the same or neighbouring trees). Changes in leaves which occur with the increasing physiological age of the tree are considered, while the longevity of individual leaves is discussed in the section which now follows.

Leaf Senescence and Abscission

A leaf typically reaches a peak of photosynthetic capacity around the time that expansion ceases, thereafter showing a gradually decline. Sooner or later a point is reached when it rather suddenly begins to turn yellow, brown, or less commonly red, and is then actively shed from the tree by completion of a separation layer in the abscission zone at its base. This change of colour is termed senescence, and it appears to be mediated by alterations to the balance of several endogenous growth substances. It is the terminal irreversible phase in the functioning of the leaf, which signals the whole sale breakdown of chlorophyll, ribonucleic acid and protein, and the rapid translocation out into the stem of some but not all of its organic and inorganic nutrients.

Abscission normally follows quickly, and the protective layer of the leaf-scar becomes an effective barries to invasion by pathogens, except for instance in cocoa where it may become a centre for *Phytopthora* infection. Lack of leaf abscission, with shrivelled leaves still retained on the stem, is often a sign that sudden stress has prevented the usual sequence of changes; retention of dead leaf basesis, however, a normal feature of some plants, including palms and tree-ferns. In *Cupressus* and *Pinus*, whole leafy branchlets areshed, rather than individual leaves, while in *Taxodium* leaves are shed from long-shootss, but the entire short-shoot is abscised.

A very wide range of factors appear to promote leaf-fall in trees,including lowered light intensity,changed temperature and photoperiod, mineral nutrient deficiency and water stress. Older leaves are much more likely to be abscised than those which have recently expanded,and it is probable that there are also interactive effects between different organs. For instance, container-grown cocoa plants subjected to drier conditions abscised a higher proportion of their leaves, and there was also evidence for an influence of the shoot/root ratio. Correlative effects were also suggested in an experiment with *Hildegardia barteri* seedings, in which, unexpectedly, more leaves were abscised under long-days than in short-days.

In the former treatment, new leaves were being produced and expanded throughout the experiment, whereas the terminal buds on plants in short photoperiods were dormant for a long time. Table 6.3 shows what appears to be a direct, additive effect of short-days and hot nights in enhancing leaf abscission in *Bombax buonopozense.* Two stages of leaf loss were detected in *Plumeria acuminata*, both of which were inhibited by light interruptions during the dark period. An interesting example is that of a disease of coffee caused by the fungus *Omphalia flavida*, which leads to premature leaf-fall and greatly reduced yields. Abscission appears to be stimulated not so much by leaf damage *per se* as by disturbance of the balance of hormones. One the flow of auxin from leaf to stem is reduced, the separation layer at the base of the petiole completes its development, and the leaf is abscised within week.

Table 6.3 Interaction of day-length and night temperature on leaf abscission in *Bombax buonopozense* seedlings. Mean number of leaves lost from each plant in a period of 4 months.

	Long-days	*Short-days*
26°C nights	0.6	1.5
36°C nights	3.0	6.0

Leaf-fall occurs all the year round in tropical forests, but peaks and troughs are usually found. The classical studies of Bray and Gorham (1964) on seasonal production of litter (about two-thrids of which was foliage) showed that in the rain forests of Colombia and Ghana, for example, most litter accumulated in March and least in July. In evergreen seasonal forest, it appears that leaf-fall may peak in the first half of the dry season, as some of the trees become completely leafless; evergreen specimens may also lose a proportion of their leaves at this time. Fluctuations in leaf-fall have many ecological implications, for instance regarding that the nutrient and water status of the soil, the photon flux densities reaching the lower tree layer, and the amount of food available to herbivores.

Every species does not behave in the same fashion of course, individual specimens in a population are often not synchronised, and conditions may vary from year to year. Nevertheless, regular annual leaf-fall has been recorded, for instance, around February for *Instia palembanica* in Malaysia. In the life of an individual tree, the partial or complete loss of its old foliage represents an important change in photosynthetic, transpiring and respiring tissue, except when leaves are produced and lost steadily throughout the year. Rather little attention has been paid until recently to the important topic of the longevity of tropical leaves.

There is no difficulty in knowing the age of a leaf on the majority of temperate zone trees: if they are deciduous species,then any leaves must be of the current year; while in many, though not all evergreens, there is distinctly 'articulate growth', such that the leaves produced in successive years are in separate 'cohorts'. In the tropics, on the other hand, it is usually impossible to know the age of a leaf unless trees are observed leaves. The minimum life-span for undamaged leaves is likely to prove to be about three months, as recorded for the woody climber *Grewia carpinifolia* in Ghana. Although the majority of leaves probably do not last more than about fifteen months, there are some well-documented cases, such as the understorey trees *Drypetes parvifolia*, *Vepris heterophylla*, and canopy species *Diospyros abyssinica* and *D. mespiliformis*, and conifers of montane forests, in which the life-span is at least 2-3 years.

There is a considerable amount of confusion in the literature over the distinction between the evergreen and deciduous habits in tropical forests. Basically, the question of whether a forest, tree or branch is leafy or leafless is a function of longevity and the relative timing of bud-break and leaf abscission, perhaps modified by herbivory and other damage. Because all shoots on a tree, all individuals of a species,and firm 'leafiness' categories becomes increasingly difficult the larger the unit. Indeed, even within a single shoot, the older leaves may be shed before those more recently expanded,while in compound leaves some of the leaflets may be abscised and other retained. Taking these points into consideration, the following four broad categories are proposed as useful groupings that are based on recognisably different growth strategies, and can be applied at any level:

A. Periodic growth (deciduous)

Leaf-fall occurs well before bud-break; life-span about 4-11 months. In this case, the branch, whole tree or forest is leafless (or nearly so) for a definite interval, varying from a few weeks to several months. Even in rain forest receiving a high rainfall evenly spread throughout the year, canopy trees of *Cordia alliodora* may for instance remain leafless for weeks in the middle of the rainy season in Costa Rica. In seasonal forests in West Africa, some specimens of *Terminalia ivorensis* can remain leafless for more than 6 months, mainly during the drier months. In this type, leaf-fall and bud-break are apparently not directly connected with each other.

B. Periodic growth (leaf-exchanging)

Leaf-fall associated with bud-break; life-span often about 12 (or 6) months. Here the flushing of new leaves starts around the time at which the majority of the old ones are absicised, with in about a week either way. This habit is clearly seen in *Terminalia catappa*, where the old leaves turn red and fall and the 'simultaneously replaced by new. In this species, the process often occurs twice a year, while in *Entandrophragma angolense*, *Dillenia indica*, *Ficus variegata* and *Parkia roxburghii*, for example, it happens once. In upland forest in Costa Rica, *Quercus* spp. can be seen to shed all the old leaves, and then within a few days to be expanding many new red leaves. It is likely that it is the senescence of the old leaves which provides the physiological signal for bud-break, placing type B responses in a different category from either deciduous or evergreen. The term 'leaf-exchanging' should be distinguished from the somewhat confusing expression 'leaf change' (*Laubwechsel*), which tends to incorporate elements of flushing, leaf growth and abscission. 'Semi-deciduous' is unsuitable because it is often used to imply that a substantial number of trees in a forest are deciduous.

C. Periodic growth (evergreen)

Leaf-fall completed well after bud-break; life-span about 7- 15 months or more. This is a truly evergreen category,which is characteristic of woody plants of the lower and middle tree layers, and some of the upper tree layer. Many dipterocarps are periodic/evergreen, and other examples include *Clusia rosea*, *Fagraea fragrans*, *Mangifera indica* and *Pinus* spp. As in stype A, there may be no direct connection between leaf-fall and bud-break,but it is possible that the expansion of new leaves might lead to senescence of the old through a hormonal control system, by competition for nutrients, or because of shading.

D. Continuous growth (evergreen)

Continual formation and loss of leaves; life-span variable from about 3-15 months. This is a second evergreen class,and is characteristic of some palms, conifers and tree-ferns, and is found in seedlings of many dicotyledonous trees. Examples of older trees include *Dillenia suffruticosa*, *Trema guineensis*, *T. micranthum*, and the shrub *Hamelia patens*. In this type there is, of course, no bud-break, but the rates of leaf initiation, enlargement and abscission may each vary considerably, according to chages in the environment, or to external and internal competitive effects. The 'leafiness' may thus fluctuate, although this was not found to be the case in the red mangrove, *Rhizophora mangle*.

The presence of type A trees even in the most uniformly moist of rain forests is an important point that is often overlooked. It provides a reminder that the microclimate in which leaves in the euphotic layer develop, particularly on the exposed crowns of emergents, is quite unlike the ever constant, equable conditions of journalistic fables, and can be quite stressful. In evergreen seasonal forests, one-third of the tree species may show the deciduous habit, and the timing of the leafless period in Ghananian trees closely followed the rainfall patterns. There were some leafless trees to be found in every months except June, but the proportion of trees without leaves increased to a peak at the height of the dry season in December and January. The percentage declines as flushing occurs in the majority of species in February and March, and falls to a very low level before the onset of the rainy season.

In evergreen seasonal forests, as in other drier formations, there seems little doubt that the chief adaptive value of the deciduous habit lies in the avoidance of severe water stress in the middly part of the dry season. Yet it should not be forgotten that many other trees operating on type B and C strategies endure the same dry seasons in a leafy state, just as various evergreen trees survive the cold winters at higher latitudes, while other species do so leafless. Not just one, but several kinds of developmental strategy commonly evolve in response to environmental or biotic problems. Moreover, drought is clearly not the only stress involved because some species lose their leaves in the rainy reason, perhaps because light intensities are lower than in the summer dry season. It should also be borne in mild that, as Janzen and Wilson (1974) have pointed out, there may be an extra 'cost' attached to being leafless in the tropics, since stored carbohydrate can be reduced by as much as 50 percent because of high respiration rates in the twigs, boles and roots.

Although the usefulness of the underlying feature of 'leafness' is beyond doubt, it has been lessened because of considerable variations in usage and precise meaning of the terms used. The frequency with which the above four categories of leaf and bud behaviour occur provides a basis for improving the physiognomic classification of tropical forests. A useful distinction has been made,for instance, between facultatively deciduous trees, such as *Ochroma pyramidalis* and *Tectona grandis*; and obligately deciduous species,such as *Ceiba pentandra*, *Cordia alliodora* and *Enterolobium cyclocarpum*.

Experimental studies on the relationships between leaf abscission and the growth and development of other organs may also be stimulated if these classes are used, for example the effects of defoliation on bud-break in type B and C shoots. Although not discussed in detail in this chapter, one of the most important consequences of all the different strategies of leaf development and maintenance will be the overall influence they have upon dry weight accumulation in the tree and forest crops. However,some assimilates will be used up in producing the organs discussed in the ramaining sections.

Cambial Activity

A further dimension to growth and development is the increase in thickness of organs through the activity of the vascular cambium. Indeed,the secondary tissues produced by this meristem constitute much the greater part of the plant body in large perennials, except for palms and bamboos. In time,individual tropical trees can become extremely large, with extensive crowns and massive root systems, and diameters at the base of the trunk (exclusive of any buttresses) exceeding 2 m. Details of some big specimens are given in Richards (1952), including an *Entandrophragma cylindricum* in Nigeria a record of a 5.4 m diameter at the base. There is also a record of a 4.1m diameter *Bertholletia* spp. in Brazil, and a 3.8 m *Balanocarpus heimii* in Malaysia. Such trees may have total dry weights of the order of 100 t, and (although not as large as some of the giant redwoods, eucalypts and *Agathis*), they rank amongst the largest living organisms in the world.

Yearly increases in main stem diameter vary greatly in different species, and from one individual to the other. As the trunk grows in size, the cross-sectional area of new tissue required for a given diameter increment gets larger. Thus, for example, what is produced by a 0.1 m diameter tree when it grows 10mm would only increase a 1 m diameter tree by about 1 mm. In middle age, trees in natural forest that have their leaves mostly in the euphotic layer may add around 5-20 mm, but in very old emergents the increment is generally lower, Amazonian rain forest, it was predicted from the stand structure that the average tree in the diameter class 0.25-0.35 m would be growing at about 8 mm a year; for larger trees the increment would decline below 4 mm. In a wet tropical forest in Costa Rica, analysis of 13-year increments of 45 tree and one liane species by growth simulation showed a range of median annual diameter increases between 0.3 mm and 13.4 mm.

Young saplings and pole-stage trees often make slow radial growth, suppressed by shade cast by the canopy trees and competition from their root systems. In gaps and along forest margins, however, the same species may be capable of rapid growth, while colonisers may be even faster growing. In plantation conditions, too, substantial annual diameter increments can be sustained; for example approximately 45 year old *Triplochiton scleroxylon* in line enrichment plantings at Mabak, Cameroon, were still producing 12 mm. Exceptionally in young, dominant individuals, values of 20-60 mm can occur, and 90 mm has even been recorded in *Ochroma lagopus.*

Although there is little problem in relating approximate average diameter increments with age in fast-growing plantations, accurate measurement of the rate and duration of radial growth is usually difficult in tropical woody plants. Increase in thickness is generally the result of activity by two different lateral meristems, the vascular cambium and the phellogen (cork cambium), both of which may be cutting off cells towards the outside of the trunk and towards the inside. Moreover, successive girth or diameter readings do not necessarily show the combined acitivity of the two cambia, even if a standard measuring position is marked on the trunks (generally 1.3 m, on the upper side of the tree if there is an appreciable slope).

In the first place, this is because there may be irregularities present such as ridges or spines,branch and leaf bases or scars, or major convolutions in outline. Secondly, bark may be abscised in various amounts and fashions,sometimes in whole sheets, as in *Eucalyptus* spp, and *Tristania* spp. of the Myrtaceae. Thirdly, tree trunks regularly shrink appreciably during a hot, sunny day, because the xylem sap is under considerable negative pressure. Accurate measurement therefore requires continuously recording dendrometeres, unless all the manual recording can be completed by 0600 h ! The existence of tension in the conducting elements of the xylem in the middle of the day can be simply demonstrated in *Dillenia* spp. and some lianes if the crown has been in sunshine for some time. If a clean cut is made into the stem with a machete, a distinct sound can often be heard as air is drawn in.

In many cases, interest centres on the major component, the porduction of xylem by the vascular cambium. Increment cores or anatomical examination are needed to provide accurate information, and a further problem arises in natural forest from the frequent lack of clear and reliable annual growth rings in many tree species.

The problem of knowing how old tropical trees are introduces inaccuracies, not just for calculating cambial growth rates, but also for every other aspect of the analysis of growth and stand structure in tropical forests.

There is some prospect of newer methods being developed,such as modelling and detailed examination of the transverse surface for small changes in chemical or physical structure. For instance, the $^{14}C/^{12}C$ ratio in carbon dioxide changed during the last 30 years, because of the atmospheric testing of nuclear weapons in the late 1950s and early 1960s, before this was banned by international agreement. This left a characteristic 'thumbprint' in the permanent structural cell components of all woody plants, which varies only slightly according to latitude. For many purposes, simpler methods may be more appropriate, such as the marking of the xylem at a particular points by dyes or damage with pins or nails, or by opening a small annual 'window'. The living cells in the wood often ramain free of starch for a considerble time after they have been formed, which could be used to determine the increment in numbers.

In order to understand more about the physiology of cambial activity, it is probably best to concentrate on species with clearcut, annual rings. There are more of these than is generally realised, including *Homalium tomentosum*, *Pterocarpus indicus* and *Tectona grandis* in South-East Asia; *Bursera simarouba*, *Goethalsia meiantha*, *Swistenia mahogani* and *Taxodium distichum* in Central America; *Ceiba pentandra*, *Chlorophora excelsa*, *Entandrophragma* spp., and Terminalia spp. in West Africa.

Most of these are deciduous or leaf-exchanging species and the former group were shown many decades ago to exhibit dormancy of the cambium. By a combination of careful measurement, experiments and anatomical studies, Coster (1927-28) demonstrated the cessation of cambial activity at the beginning of the leafless period, with divisions occurring again when flushing took place. He also showed that the stimulus for renewed cambial acitivity in leafless plants must arise from the expanding buds, since the cambium remained dormant if they were removed, or if they were isolated from the portion of cambium under study by complete ringing. The influence of the expanding buds was not exerted through photosynthesis, for the cambium was still activated if the plants were kept in the dark, so Coster proposed a hormonal explanation at a time when plant growth substances had only recently been discovered.

Thus it is likely that auxin,moving basipetally in the phloem from expanding buds, is primarily responsible for initiating cambial activity. Other hormones are probably involved as well, since for instance the cambium in dormant, leafless teak shoots did not respond to IAA unless bud dormancy was artificially broken with ethylene chlorhydrin. After leaf expansion has been completed, it is generally considered that there may be sufficient auxin produced by the mature leaves for cambial activity to be maintained, but when they all senesce the supply dwindles and the cambium becomes dormant. Indeed, Coster was able to cause the cessation of cambial activity by ringing the stem between the mature leaves and the part of the cambium being studied.

Conversely, there is recent evidence that the living cells of the wood may influence primary shoot growth. The vertical parenchyma cells, especially those associated with the vessels show high levels of acid phosphatase enzymes, particularly in the pits linking the two types of cells. It seems likely that sugars are being secreted into the vessles, as occurs in the temperate zone sugar maple tree, but for much longer periods, and in more species in the tropics.

The timing of the periods of cambial activity and rest is thus likely to be controlled indirectly by the same external and internal factors which determine bud-break and leaf-fall. One exception is probably provided by trees that flower during the leafless period, where opening floral buds and developing fruits could provide the stimulus for an earlier resumption of cambial activity. This is one of the many possible reason for 'flase' rings in the xylem, which mean that not every deciduous tree shows clear-cut annual growth patterns in its wood.

In the leaf-exchanging trees of type B, which are leafless for at most a few days, the cambium may not become fully dormant. Indistinct rings were formed in *Khaya grandifoliola*, but clear annual rings in *Dillenia indica* and *Ficus variegata*. In evergreen trees of type C, continuous cambial activity is probably the rule, although a few samples of cambial dormancy and even annual rings have been reported. More generally, they are indistinct, irregular in occurrence or absent, espcially when seasonality of climate is slight. The clear-cut growth rings in *Avicennia germinans* consist of alternating bands of wood and softer phloem-containing tissue, and do not indicate age. The most vigorous shoots can contain up to 5 rings, all of uniform width. In evergreen trees of type D, including young saplings that are continuously expanding new leaves, the cambium is almost certainly active throughout the year.

Whether cambial activity is continuous or periodic, there are often large fluctuations in the rate at which cells are formed and differentiated. As generally recognized, there is a clear positive correlation between the rate of diameter increment and the amount of height growth being made by a tree, and there is often an even closer link with its crown dimensions. Cambial activity may in some cases by most rapid which shoots are extending, but in cocoa it was slow while flushing proceeded, and reached a peak during the rainy rseason,. Similar fluctuation has been recorded for other tropical tree species.

The factors which control the rate of cambial activity have been little studied experimentally, in spite of their obvious importance of forestry. Some external influences may be expected to act directly on the cambial initials or the differentiating cells, as for example temperature and moisture stress, or the effect of gravity in branches and leaning main stems. Others, for example photoperiod and mineral nutrition, probably affect the cambial zone indirectly, perhaps by modifying growth or internal control systems, such as the partitioning of assimilates between various active meristems or into storage sites. High temperatures between 35°C and 40°C, which reduced the rate of shoot extension in *Ceiba pentandra* compared with somewhat cooler conditions,also significantly depressed diameter increments. Water stress was implicated in a study of cocoa, in which plants supplied with the equivalent of the annual rainfall (1,830 mm) spread evenly over the whole year made about 30 percent more diameter growth than those receiving natural precipitation.

Considerable genetic differences within species in their diameter increment have frequently been reported between provenances of tropical forest trees. Pronounced clonal variation also occurs.: for example, both the radial increment of xylem and the number of cells produced radiallly was more than twice as great in one clone of *Terminalia superba* than in another. In terms of cross-sectional are (basal area) or volume production of wood,such differences become even larger, since the calculations involve squaring the data.

Significant clonal differences also occurred in *T. superba* in the radial dimensions of vessels and of fibres + parenchyma cells, which may be expected to alter the wood properties. These will clearly be affected as well by the proportion of the different types of cell that are produced. However, although the patterns of different tissues have been extensively described, and are often of value in identification of timbers, there is as yet little understanding of what determines the type and size of cell produced. How, for instance,do the narrow,tangential bands of parenchyma and tracheids which mark cessation of cambial acitivity at a growth ring differ from those which are often formed during continuing wood production? Scanning electron microscopy in association with detailed studies on clonal material growing in controlled environments might clarify this. Moreover, there is still much to be done in relating the structure of wood to its properties, and here the work of botanists and those in the timber trade may be assisted and integrated by a useful handbook by Chudnoff (1984), which covers the extremely confusing area of local and botanical names for 374 different timbers, as well as many aspects of their utilization.

Root Formation and Development

The first roots to be formed in a young seedling are usually a group of first-order laterals that arise from the tap-root just below ground level. Like most roots, these do not grow vertically downwards but elongate more or less horizontally. If the tip of the primary root of young cocoa seedlings is decapitated, some of the laterals nearest to the cut curve downwards and become positively geotropic. Thus, even at this very early stage, it is clear that there are dominance relationships within the root system, and these presumably lie behind the various characteristic organizational patterns found in older trees. Because of the difficulties involved in following individual subterranean roots, the majority of work on branching has been done using aerial roots. Unless damaged, these typically show little or no branching while in the air, but lateral roots start to appear if the tip makes contact with the ground. In *Cissus*, branching is induced if the root is placed in water, but in *Rhizophora* darkness appears to be the chief stimulus.

The initiation of aerial roots takes place within the shoot system, providing an opportunity to examine the formation of new root apices above-ground. Trees with stillroots, stranglers and epiphytes may all provide potential material, while climbers such as *Cissus aralioides* and *Urera* spp., actually spread vegetatively through the formation of adventitious roots. *Anthonotha macrophylla, Scaphopetalum amoenum* and *Sleotiopsis* usambarensis form new roots during the natural layering of their drooping branches; while conversely the rarer case of the formation of shoot apices on root tissue could be studied in plants forming root suckers, such as *Bumelia reclinata, Cedrela serrata, Chlorophora excelsa, Cordia alliodora, Drypetes diversifolia, Melia azedarach, Millettia thonningii* and *Trema* spp.

Despite these and other instances, vegetative spread is thought tò be relatively uncommon amongst the woody plants of the tropical forest. However, the capacity to produce roots in stem cuttings appears to be widespread; particularly if the material originates from young trees and so perhaps natural vegetative reproduction may turn out to be commoner than expected. Since many commercially important plantation species can be rooted from stem cuttings, this has a number of important implications for research, tree improvement and clonal forestry.

It also provides favourable circumstances for studying the formation and out growth of roots, since large numbers of cuttings can be set and removed at intervals from relatively uniform propagation beds. Few anatomical investigations have been done on tropical trees, but there is a growing literature on environmental, harmonal and other effects on root initiation. For instance, rooting of cuttings of *Triplochiton scleroxylon* under automatic mist was promoted when the temperature of the propagation medium was raised to 25°-30°C, compared with 20°C and these conditions appear to be suitable for many other genera.

There are often substantial differences in rooting ability between clones; and also amongst cuttings that are of different sizes,originate from different parts of a tree,or from tree pretreated in various ways. The addition of auxins may increase the rate of rooting, and the number of root initials formed,but when supra-optimal can inhibit root elongation. Rooting with and without auxin in the medium has even been achieved with teak in aseptitic culture,using small shoots produced by meristem proliferation.

The daily course of root elongation was followed in aerial roots of a large liane, *Cissus adnata*, whole foliage was high in the forest canopy. The roots showed little or no growth when the sun was shining or when a drying wind blew during the night, but immediately after rain the rate of elongation rose temporarily to the very high value of 30 mm/h or more. Coster showed that the chief factor involved was the water status of the plant, and found that the roots could actually shrink by as much as 7 percent in length when subjected to drying. In contrast, the aerial roots of *C. sicyoides* grew at a constant 4 mm/h throughout the night and a sunny day.

Seasonal changes in the rate of root growth no doubt occur widely, but have been little studied. Nor is it clear whether there is any general cessation of root acitivity at certain times of year, as there usually is in the shoot system. Only in extreme cases does temperature appear to stop elongation : *Citrus* roots for instance grow only above 10°C. However, it may well be that root and shoot growth are linked by internal control systems as well as being influenced by their separate external environments. For instance, in controlled environments the root and shoot growth of cocoa are both rhythmic, with peaks and troughs of acitivity alternating. The rapidly expanding leaves of the flush apparently inhibit the roots, because if the foliage was all removed immediately after unfolding, the roots showed a high constant growth rate during this period.

Coster (1932) grew about 70 tropical (mainly woody)species at a site in Java which had a deep, permeable soil and regular rainfall. After 6 months, he dug out the seedling root systems carefully and found that in general the main root was longer than the main stem, and the spread of the horizontal surface roots was greater than that of the crown. Average root elongation rates were over 20mm per day for the most rapidly growing trees, which is faster than that found in most temperate trees. The total length of the main root plus the primary or first-order lateral roots had actually exceeded 30 m in *Melia azedarach* and *Sesbania sesban*, though others had grown much more slowly.

Just how extensive root systems can become is vividly illustrated by the fact that coffee plants can produce roots totalling 25 km in length by the time they are three years old. At La Selva, Costa Rica, it was estimated that more than 250 gm^{-2} dry weight of fine roots was produced in the first year by woody regrowth. This was roughly comparable to that present in the 5-year-old secondary forest that had been cleared, indicating a very rapid restoration of the nutrient uptake capacity. Fine roots often have a short life-span: it took much longer for the new structural root systems to develop. A major problem that hampers studies of the dynamics and turnover of root systems is the long, hard and tedious work involved indigging out, separating and assessing the different types of root.

It is seldom possible to attempt sufficient replicates to assign variation reliably to any factor,but one exception is provided by a study of the maximum rooting depth in tea. This indicates a significant difference between 8 clones, the deepest growing of which had reached 2½ times the depth of the shallowest in the same time. Cambial activity in roots is another area which has been little studied. When aerial roots of *Rhizophora mangle* reach the ground, their cxtraordinary rapid extension rates slow, and they cease to possess an elongating zone averaging 12 mm in length. Soon afterwards cambial acitivity becomes very marked and the roots thicken.

An unusual feature is the joining together through anastamoses and secondary thickening of two roots of the same, tree termed self-grafting by Ng (1975), and noted in several species in addition to *Ficus* spp. including *Shorea leprosula* and *Swietenia macrophylla*. Grafting between different trees was found in *Gmelina arborea* and *Shorea curtisli*, and Leroy Deval (1974) considers that dominant trees in stands of the gregarious species *Aucoumea klaineana* acutally 'annex' the root-systems of dominated trees to which they are fused.

Physiological Changes with Age

One of the distinctive features of long-lived perennial plants is that they usually show variation in structure and function as they grow older. Some of the changes which occur are gradual, while others appear suddely, but in either case shoots which are collected from the top of an emergent tree seldom resemble closely those produced by the same species as seedlings in the undergrowth. The leaf of *Celtis mildbraedii* from the canopy is a quarter of the length and only 9 percent of the area of that taken from the under growth, while in the liane *Hugonia platysepala* the canopy leaf is 40 percent and 30 percent of the shaded one. Sometimes leaf pairs of this kind are so unlike each other that most botanists would assume that they belonged to two different species or genera. It is one thing to recognize that such within-tree variation occurs, but quite another to determine its cause. The leaves could be dissimilar for a large number of reasons that can be considered in four broad groupings:

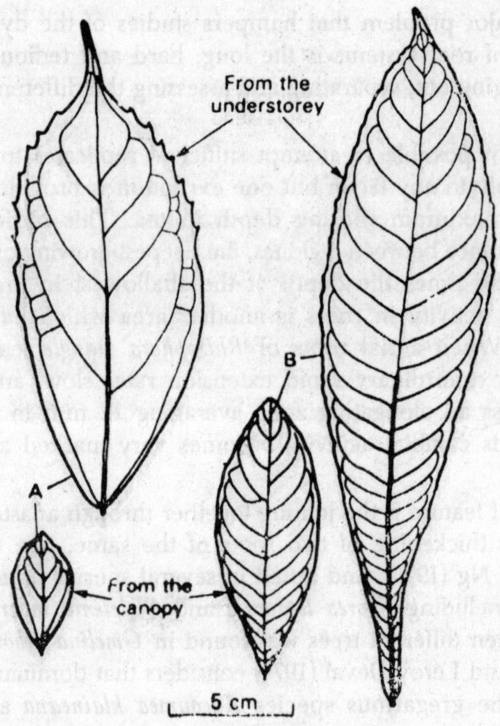

Fig. 6.7. Leaves from the canopy and the understorey of Ghanaian forest: A–Celtis mildbraedii, a large tree; and B–Hugonia platysepala, a liane.

Firstly, they have grown in very differing environments, so that factors such as light intensity and quality, temperature and moisture stress may have modified them, directly or indirectly. Although it is difficult for logistic reasons to carry out experiments using emergents, it is possible in plots of saplings or batches of potted seedlings to investigate how much their leaf size and shape very in response to treatment differences.

Secondly, they are normally bound to be of a different genotype. The extent of genetic variation is often great, especially in outbreeding species of forest trees that are 'undomesticated'. It is well known that this can influence many aspects of growth, and thus affect leaf development directly and indirectly. With the increasing use of vegetative propagation in forest tree research and for clonal foresty, the strongly inherent characteristics of clones are likely to become as easily recognized as in *Hibiscus* or

Bougainvillea, citrus or mango. For instance, potted plants of *Triplochiton scleroxylon* clone 8036 could be readily distinguished from other clones by the pronounced secondary lobing of the leaves.

Thirdly, the leaves are borne on trees of a very different size, with much greater internal competition between potential sites for growth and storage in the big tree (*i.e.*, many more 'sinks'), and longer distances for translocation. This question has been approached by experimental disbudding, using the short-lived woody plant cassava, with its relatively simple crown. This follows Leeuwenberg's model with regular and repeated increase in the number of actively growing shoot species, and associated decline in their vigour of growth. Since removal of all but one of the growing points more than doubled the rate of growth of a 'singled' apex by the end of the first week, and the leaves produced later were as much as 35 percent longer than in unpruned controls, internal competition can indeed influence leaf size. Differences in this catetgory,which are due to ageing as a tree gets older., can typically be reversed by appropriate treatment.

Fourthly, the variation between the leaves may arise because of phase-chage. These changes also occur with age, so that shoots originating from an older crown often exhibit more permanent differences from those taken from young seedlings. The change from the 'juvenile' to the 'mature' phase is termed maturation, and the features are still exhibited when the mature tissue has been vegetatively propagated and the two phases are growing is similar-sized plants in the same environment. Ideally, investigation of phase change should be done with material that is also of the same genotype, which is in fact possible if young coppice shoots can be obtained, as these are generally regarded as still being juvenile. When such detailed investigations are not possible, a reasonable compromise might be to compare the crown foliage of several large trees with leaves taken from regeneration in a large gap or clearing nearby, where (away from the margins) the microclimate is relatively similar to that at the top of the canopy.

Differences that were clearly due to phase-change were found in similar-sized mature grafts and juivenile seedlings of *Cedrela odorata* that were grown in controlled environments under different photoperiods. The rate of shoot elongation was much more strongly influenced by day-length in the mature plants, which also ceased extension for a time in both treatments. In contrast, the terminal buds of the juvenile trees only became dormant under short-days. On the other hand, differences which are definitely not due to calender age are shown in, since all the trees in the photograph are 5 years old. Because of the favourable external environment, one of the trees has far outgrown the others and has entered the category A phase with periodic growth, a rapid flush of height growth and a clear-cut deciduous period. The small trees remained ever-green, and grew continuously or for long periods, though at a very slow rate.

In the uncontrolled setting of the tropical forest, it is generally difficult or impossible to separate environmental and genetic factors on the one hand from physiological changes with age (ageing and maturation) on the other. Differences of various kinds between the foliage of large and small trees are often reported, the leaves of the former generally being smaller with drip-tips generally absent. Compound leaves often have fewer leaflets, as in the Meliaceae, for example. In some instances, the leaf shape, and sometimes the phyllotaxy, may alter strikingly with age; for example young seedlings of *Acacia mangium* have the normal pinnate leaves typical of the Leguminosae, but these quickly give way to phyllodes, with a few intervening nodes bearing transitional structures. *Araucaria cunninghamii*, *Triphyophyllum peltatum* and some *Eucalyptus* spp. have at least three contrasting types of foliage.

As trees become larger, they often cease producing leaves continuously, and the leaf-exchanging or deciduous habits may appear. In genera which show periodic growth from the start, such as *Agathis*, *Barringtonia* and *Mangifera*, longer periods of terminal bud dormancy may be found. In addition, there may be a tendency for individuals of a species to become more synchronized in their development. For instance, the typically out-of-phase condition of a 2-3-year-old plantation of *Terminalia ivorensis* contrasts with the comparative regularity of a 20-year-old stand.

As a seedling grows into a large tree, the characteristic branching patterns also emerge. Sometimes branches start to grow almost at once, while in other species there can be a period of a few months or years without any. In young saplings of some species in the Meliaceae, the rachis of large compound leaves borne on the main stem acts instead of a branch in supporting the photosynthetic surfaces. Palms that follow Corner's model, such as *Elaeis guineensis* and *Cocos nucifera*, produce very large fronds from their trunks, the only branches being reproductive and determinate. Conversely, the terminal growing point of the main stem loses its meristematic nature and disappears in cocoa, signalling the production of a 'jorquette' of plagiotropically growing branches. Neither branches nor main axis are produced during the 'grass stage' of *Pinus merkusii*, most of the growth of the young tree going into the development of the root system. Only later do the main stem and branches develop as in other pines. The grass stage is thought to be an adaptation by which the terminal bud is protected from 'light' fires by the cluster of long needles; it occurs in this species in Vietnam, but not in Sumatra.

Some interesting theoretical modelling of branching is reviewed by Fisher (1984), but the amount of experimental work on branch development has not been extensive. The formation of the first primary branches was delayed by more than 3 months, and no second-order branches were formed, when *Rauvolfia vomitoria* seedlings were grown under 8 h rather than 12 h day-lengths. Applied gibberellin promoted the elongation of both orthotropic and plagiotropic branches in cocoa, but did not change their phyllotaxy or geotropic orientation. *Triplochiton scleroxylon* clones differed strongly in the number of branches per metre of main stem and in their thickness and persistence after they had become shaded by later growth. Some clones showed a tendency to produce 'multistems' (several strong vertical shoots from near the ground). Decapitation studies with small potted plants have shown that clones vary in the number of buds which are released from apical dominance, and that mineral nutrition and gibberellins, for example, can affect the re-imposition of dominance by the uppermost lateral. Experiments with *Terminalia ivorensis* seedlings involving removal of terminal or lateral shoots showed that branches can affect the growth of the leading shoot, as well as *vice versa*.

Besides environmental and genetic effects, there appear to be various endogenous and ontogenetic factors at work : for instance, a clone of rubber trees usually formed the first branches after the terminal bud had flushed nine times. In some tropical trees there are two buds in each axil, and typically these behave differently. For example, the upper one in *Nauclea diderrichii* produces plagiotropic branches, while the accessory bud may produce an orthotropic shoot after damage or decapitation; and these differences persist incuttings taken from the two types of shoot. There can be several accessory or serial by treatment with various growth regulators, including the cytokinin benxyladenine, the anti-auxin tri-ioidobenzoic acid and compounds releasing the gaseous hormone ethylene in the tissues. The natural or induced sprouts from these accessory bud produces plagiogropic, flowering shoots. However, the upper three buds can form flowers sequentially, but the fourth is often vegetative.

In many ways, the most fascinating change in the life of a tree is the onset of reproducutive activity. Some species, notably pioneers, start flowering profusely after about 2-5 years, although the great majority remain purely vegetative for much longer than this. Nevertheless, specific precocious individuals are often noted in the literature as producing flowers before they are two years old, for example. *Anthocephalus chinensis*, *Dipterocarpus oblongifolius*, *Citrus* spp., *Funtumia africana*, *Hildergardia barteri*, *Monodora tenuifolia*, *Pinus caribaea* var, *hondurensis*, *Trema guineensis*, and many mangrove species. If such precocity is primarily genetic in origin, clones made from early-flowering individuals might provide an elegant research tool for miniaturising the study of flowering. for commercial plantations, although early flowering may be an advantage in fruit trees it is probably undesirable in species that are grown for vegetative yields. There is some evidence that there has been unconscious negative selection in forestry, perhaps because seed is more conveniently collected from precocious and prolific individuals or stands. In teak, in West Africa, for example, entire young plantations may be seen flowering or fruiting copiously and occasional individuals may even do so in the nursery when only 3 months old. Since the inflorescences are large, and terminate the main stem, this may be expected to cause poor form and reduced vegetative growth.

In most forest trees, however, there is a more extended juvenile period through which they pass before flowering commences. It is common for this interval to average between 10 and 30 years, although there is considerble variation between sites and amongst individuals. In general, it tends to be shorter in plantations than in natural forest, probably for a combination of genetic and environmental reasons; and it is also typically a briefer stage for pioneer species than for climax species. Thus Ashton (1969), for example, considers that it is often about 60 years before majority of the dipterocarps of the upper canopy of Malaysian forests start to flower. In one group of plants the onset of flowering is of particular significance–monocarpic species which flower once and then die.

Agave is a well-known example, in which the life-cylce is about 10 years, while in *Tachigalia versicolor* it ends after many years. Thus monocarpy is not confined to species such as *Corypha* palms, which exhibit Corner's model, but can occur in emergent trees with many shoot apices. Another striking feature of this group is that they often flower gregariously and since clones that have been dispersed to different regions still tend to flower and die at the same time, some long-term endogenous 'calendar' is implied. Some but not all bamboos are monocarpic, with the life-cycles often rather clearly known, ranging from a few years in *Chusquea tonduzii* to 80 years in *Rhipidocladum pittieri*. Such very long juvenile periods are thought to have evolved as extreme instances of deprivation followed by saturation of seed predators.

More experimental studies are needed on the various physiological factors which may be involved in controlling the length of time a tree produces only vegetative shoots, particularly because it is often desirable to be able to shorten this in crop species. However, it is known that coffee is a short-day plant in its floral initiation, whereas *Stylosanthes guianensis* is a long/short day plant. In *Bougainvilleo* spp, and *Euphorbia pulcherrima*, the first flower production in these vegetatively propagated ornamentals is affected by an interaction between photoperiod and temperature.

Cuttings originating from young seedlings of *Triplochiton scleroxylon* have flowered on several occassions under warm, long-day conditions in glasshouses in Scotland, at an age of 2-6 years from germination. Observations on the nut-bearing *Shorea penanga* in a nutritional trial in Sarawak suggested that profuse flowering and fruiting at 6 years occurred particularly in plots receiving all the fertilizers. Immature mango shoots have also been induced to flower and fruit by treating them with potassium nitrate.

The most generally applicable treatment for stimulating forest trees to flower may prove to be growing them as quickly as possible, which appears to hasten the processes involved in maturation. However, various additional techniques may be needed once the trees have reached 'ripeness-to-flower', to provide them with the 'opportunity-to-flower'. Methods which have had some effect in seedlings include removal of the terminal bud in mango, decapitation in coffee just after the first branches are formed, and defoliation in guava. Bark-ringing has also been successfully used in a number of tropical as well as temperate woody plants.

Some indication of possible hormonal effects is suggested by the induction of flowering in some 10 year old cocoa trees, that had never been seen to flower previously, by transplanting into their bark small clusters of flower buds from very floriferous trees. Treatment with gibberellins stimulates cone initiation in young trees in the Cupressaceae, and the techniques may be applicable to other conifers. Other plant growth regulators may also be invloved; for example compounds which release ethylene in the tissues have stimulated the onset of flowering in mango. The herb *Geophila renaris* remains vegetative as long as the soil moisture content is near its maximum value. If it falls below the permanent wilting percentage flowering is stimulated, and it continues afterwards even if the soil again becomes saturated with water.

Flowering

Flowers often attract the attention, of people as well as of potential pollinators. However, many open only for short periods, and may be hard to see high up in the euphotic zone. There is an extraordinary range of shapes, sizes and distribution of flowers in the tropical forest with diverse colours, markings and often scents. *Raffesia arnoldi* forms a single, foulsmelling flower about a metre in diameter; cocoa bears undreds of small, cauliflorous blossoms; *Bombax buonopozense* unfolds red, waxy 'ponds' high up in the canopy, in which aquatic invertebrates and vertebrates can flourish; while a large dipterocarp tree might remain purely vegetative for many years, and then suddenly produce a million flowers that all open within a week or two. With diversity on this scale, it is obviously difficult to generalise, although fortunately there is now a growing literature on floral biology at the community and species levels.

The critical first stage in reproduction is the conversion of an apex from producing only vegetative primordia to forming an inflorescence, cone or single flower instead of or in addition to foliage leaves. The first detectable signs are often an alteration in the shape of the apical dome, and in the growth rates of leaf and bud primordia, as well as inhibition of the sub-apical meristems which produce and extend the internodes. Floral development may continue without a break from initiation until the opening of the flowers, as for example in *Hibiscus* and herbaceous species. However, in many trees there is an interval during which the initials remain within a dormant flower bud, which can vary from a few weeks to many months, or even 2½ years in some *Eucalyptus* spp. It is therefore necessary to distinguish carefully between the timing of floral initiation, and that of flower opening, since the factors which influence these two processes may not be the same.

More or less *continuous* flowering is found in some evergrowing secondary tree species such as *Dillenia suffruticosa, Harungana madagascariensis,* and *Trema orientalis.* The common mangrove *Rhizophora mangle* also flowers continuously, although in Southern Florida it shows a peak in June/July. In the great majority of cases, however, the flowers emerge *periodically,* either regularly or irregularly, though there is usually sufficient overlap for the forest as a whole never to be without flowers. Individual trees may also show variation within the crown, with some parts in flower, other branches perhaps in fruit, and others again producing only leaves.

On the other hand, synchronized flowering can occur over many hectares of forest, as in *Pterocarpus indicus* in Malaysia, *Tabebuia serratifolia* in Surinam and theWest African lianc *Calycobolus heudelotii.* A few species that flower annually are so reliable in their timing that they have been used as a signal for the planting of a food crop–for example., *Erythrina orientalis* for yams in the New Hebrides, *Sandoricum koetjape* for rice in West Malaysia and *Trichilia heudelotii* for the second planting of maize in Ghana. Perhaps they have proved more useful than astronomically based calendars, since they co-ordinate the agriculture cycle with current environmental conditions rather than with fixed dates, which do not allow for year-to-year variation in temperature and precipitation.

The frequency with which flowering periods recur varies from about 3-4 months in *Ficus sumatrana,* for example, to 10-15 years in *Homalium grandiflorum* Guava can flower twice a year, while fairly regular annual flowering occurs in the fruit trees *Baccaurea parviflora* and *Xerospermum intermedium* and in a number of common forestry plantation species, including *Cedrela odorta, Gmelina arborea, Tectona grandis, Terminalia ivorensis* and *Pinus kesiya.* Biennial tendencies appear to be characteristic of most species reaching the canopy in Surinam forests and may also become marked in older mango trees. Although dipterocarps may occasionally flower again after 1-2 years, the interval is more commonly about 3-8 years. In these irregularly spaced 'mast' flowering and fruiting seasons, flower profusely and gregariously over a period of a few months, often accompanied by heavier than usual reproductive acitivity by trees in other families.

In a detailed study of six related *Shorea* spp, which involved frequent assessments high up in the canopy, Chan (1977) found that each species occupied a specific 'time-niche', with the flowers open for about 15 days in the first to flower and for about 25 days in the last. Peak flowering times of the 6 species did not overlap at all. This remarkable characteristic, which has also been found in the New World tropics for 4 *Guarea* spp. and in the Bignoniaceae is thought to have evolved through competition for a limited number of pollinators. There is considerable speculation about the nature of the physiological trigger for irregualrly spaced, but community-wide flower initiation. Perhaps the most likely hypothesis is that an increase in the diurnal fluctuation of temperature, associated with drier spells of weather with clearer skies, stimulates floral initiation at the same time in all the six *Shorea* spp. which then vary in their rates of subsequent flower development.

Because some tree species show reproductive activity that is clearly or probably related with seasonal changes does not mean that all them do so. The proportion of 'net clearly seasonal' species was 14 percent in south Florida, 26 percent in a detailed study at Barro Colorado Island, Panama, that also included herbaceous plants, and about 33 percent in Ghana. Some examples of a seasonal tree species are *Anisophyllea corneri*, *Anthonotha macrophylla*, *Blighia sapida*, *Genipa caruto*, *Santiria laevigata* and *Trophis racemosa*.

Considering the forest as a whole, however, there are usually peaks of flowering at certain times of year, even in rather uniform climates. Comparisons between data from different regions can be made using a *seasonal flowering index*, which is the ratio between the number of species flowering in the 6 month period with the most flowering, and that with the least. Values between 1.2 and 4.9 have been calculated for consecutive 6 month periods, the degree of seasonality depending both on the magnitude of the difference and the amount by which the number of flowering species exceeds zero. A seasonality at the community level has been claimed for a forest in West Malaysia, but it is interesting that if the data presented are recalculated, there is 25 percent more flowering in the months March-May and September-November than in June- August and

December-February. This difference is statistically significant, and gives a seasonal flowering index (on a 3 month + 3 month basis) of 1.25.

Not surprisingly, there tend to be closer links between reproductive phenology and particular times of year in forests experiencing definite climatic seasons. Very often the flowering peaks occur during the dry season and the early part of the rains. Deciduous species tend to flower in the dry season, expanding flower buds that do not also contain foliage leaves (e.g. *Ceiba*, *Hildegardia*), or mixed floral and leaf buds at the usual time of vegetative bud-break. Peak flowering in evergreen species occurred during the transition to the wet season, the inflorescences or single flower often emerging with the new leaves, or perhaps being produced cauliflorously.

There is more variation found between different forests when climate is relatively uniform. Most flowering occurred from March to July in a six year study of 56 individuals (42 species) in West Malaysia, whereas June and July were not good flowering months in a four-year study of 131 trees (62 species) only about 50km away. At Pasoh Forest Reserve, only about 150 km away, a 3½ year study of over 400 trees (19 fruit tree species) showed a clear predominance of flowering in the earlier part of the year. A comprehensive study of the forests of Panama, invloving three years fieldwork and the examination of more than 50000 herbarium specimens, suggests that the different life-forms do not all flower at the same time. For instance, trees, lianes and epiphytes showed peak flowering in the dry season, whereas shrubs and herbs growing in the undergrowth or in small gaps flowered especially in the rainy season.

The *timing of floral initiation* has been little investigated, although it is clearly an important topic that is relatively easy to study by dissection if suitable material can be collected regularly. This would be easier in species such as *Ateramnus lucidus*, *Bosqueia anagolensis*, *Monodora tenuifolia* and members of the Cupressaceae, in which reproductive and vegetative shoots can be distinguished several months before flower opening occurs, because the floral buds are larger or the developing cones appear different. Besides clarifying the questions of seasonality of flowering discussed above, knowledge of when and where flower formation occurs in a tree can greatly assist physiological studies into the factors which promote floral initiation. For instance, if it is known that terminal or lateral reproductive organs start to form just as shoot elongation ceases, experimental treatment can be applied with much greater temporal and spatial precision.

Coffee, tea, bougainvillea and poinsettia are all facultative or obligate short-day piants, and the suppression of flowering in *Hildegardia barteri* growing close to street lights has been interpreted in the same way. However, it should not be concluded that all tropical trees will be short-day plants or day-neutral, for a substantial minority of tropical plants have been reported to be long-day species, while temperature can modify effects of photoperiod. Drought stress has long been claimed to be a factor stimulating flower initiation, although the evidenceis often lacking for tropical trees.

The most effective hormonal treatment for stimulating reproductive activity is undoubtedly the application of gibberellin (GA_3) to members of the Cupressaceae. Large numbers of male and female cones can be stimulated, particularly if doses of about 1-100 mg are injected in alcohol into small holes gibberellins ($GA_{4/7}$) appear to be promising tools for promoting cone initiation in *Pinus* spp. but in broad leaved trees it appears that gibberellins often inhibit floral initiation. Three different growth retardants promoted the formation of flowers in mango during an 'off' year, especially in combination with bark-ringing. Pronounced stimulation of flowering following a soil drench application of another inhibitor was found in the vine *Clerodendrum thomsonae.*

Other treatments which have been found to modify flower initiation include bark-ringing,which is stimulatory in *Faurea speciosa, Lonchocarpus utilis, Pinus elliottii* var. *elliottii, Terminalia ivorensis,* mango and rubber. It may also be possible to modify the sex of inflorescences of *Elaeis guineensis* towards male by removing the older leaves, although there is a lapse of two years before this is seen because of the long period of development after initiation. In *Carica papaya,* which is dioecious, there is some evidence that male trees can produce functional female flowers if the stem has been damaged.

Progress in the physiological study of flower initiation is likely to come from a sustained effort to develop miniaturised systems in which flowering can be studied in depth. In particular, the use of selected clones that can initiate flowers under controlled environments could transform our knowledge of the effects and interactions of external factors, and also elucidate the internal links and balances between the formation and growth of flowers, leaves, stem units, roots, etc. There is even the prospect nowadays of studying reproduction in aseptic cultures, in which even the hormonal balance of an organ can be carefully controlled.

The *opening of flowers* has been studied extensively in coffee, and also in trees and shrubs in a Costa Rican forest with a pronounced dry season. *Bernardia nicaraguensis* and *Croton reflexifolius* initiate and develop floral buds, and then become leafless during the dry season. Just 2-5 days after the first rains, they flower gregariously, whether this happens to be in March, April or May. Other woody species also appeared to be triggered by rainfall to open their flowers, taking longer to respond, but tending to occur each year in the same sequence. The shrub *Hybanthus prunifolius* similarly flowers synchronously is Barro Colorado Island, Panama, a few days after the first rainfall of 12 mm or more,provided that the preceding dry period has been sufficiently intense and prolonged. Watering a plant after dry weather will stimulate it to flower, but watering it during the dry season prevents it from flowering.

A rapid drop in temperature normally accompanies a sudden tropical thunderstorm, and in coffee either a temperature-drop of at least 3°C in 45 minutes, or the relieving of water stress by the rain, can trigger flower opening after a dry spell. In the epiphytic orchid *Dendrobium crumenatum*, the change in temperature appears to be the critical factor; whereas with drybulbs of the 'rain-flowers' *Pancratium* and *Zephyranthes*, which flower 3 days after heavy rain, a direct effect or moisture is indicated by their reponse to either warm or cool water. The environmental factors and changes in endogenous promoters and inhibitors that are associated with coffee flower opening are discussed by Browning (1977).

A suprisingly large proportion of flower buds may abort and be abscised before opening, particularly when there are large numbers present, as in *Shorea* spp, and *Triplochiton scleroxylon*. Sometimes the flowers on a plant all open within a week or so, as in the gregraious examples already quoted. In other species, such as *Jacaratia dolichaula* and *Pentagonia macrophylla*, a few flower buds may continue to open for a long for a long as 3-4 months. Individual flowers may last for days, as in mangroves or only for hours, as in many other tree species, and they frequently have very precise opening times that are clearly related to the behaviour of the appropriate pollinator. Thus, for example, anthesis starts around 1600 h and is completed by 2000h in *Durio zibethinus*, while pollination occurs mainly between 200 and 0100 h. The flowers of *Cieba pentandra* open as dusk falls, and are pollinated by bats that take the nectar and distribute pollen on their fur.

Many other physiological and structural aspects of flowers and pollination are covered by Tomlinson (1980), Bawa (1982), Tanner (1982) and Baker *et.al* (1983), amongst others. An important point is that around 30-40 percent of tropical trees have been found to produce separate male and female flowers, the majority of these being dioecious, with separate male and female trees. Various intermediate categories also occur quite frequently, serving as a reminder of the great variability in reproductive biology which exists in the tropical forest.

Fruit Development

Except in parthenocarpic plants such as banana, plantain and seedless varieties of *Citrus*, pollination must take place if fruit-set is to occur. Unpollinated flowers are usually abscised, and it is common for this to occur to many pollinated flowers as well. High proportions of flower abscission have been found for instance in *Aucomea klaineana* and *Shorea* spp. Indeed, there can be several thousand flowers on a single inflorescence of *Parkia clappertoniana* or mango, and yet only 1-5 of them normally set fruit.

One reason for this may be that flowers which are fertilized first often inhibit fruit-set in those that are pollinated later. On the other hand, in species such as epiphytic orchids that produce relatively few flowers, it may be noticeable that the individual flowers appear to set fruit independently of each other. External factors can also affect the setting of fruits : for example, water stress and insect attack are both likely to increase flower abscission. Either high or low temperatures partially or completely inhibited fruit-set in coffee, the optima being 23°C day/17°C night, or 26°C day/20°C night.

The subsequent enlargement and development of fruits takes place at rates that are characteristic of the species, modified both by the prevailing environmental conditions and by mutual competition between fruits and with other 'sinks'. By no means every fruit that is set reaches maturity, partly because fruitlet abscission continues to 'thin out' the crop remaining on the tree, even if there is no abnormal stress. Insect predators or mammalian herbivores may damage seeds or eat fruits before thay are ripe, while diseases such as cherelle wilt in cocoa may cause a high proportion of the young fruits to shrivel on the tree.

The final stages in fruit enlargement are often relatively slow, but then ripening can occur quite rapidly. During the latter process, a sudden rise in respiratory rates can frequently be observed in fleshy fruits, associated with the metabolic be observed in fleshy fruits, associated with the metabolic changes leading to softening and the development of the characteristic colour and flavour. The onset of ripening may be stimulated by an increase in ethylene levels within the fruit, associated with changes in auxins. In avocado for instance it has been shown that low levels of applied indoleacetic acid suppressed ethylene production and delayed ripening, whereas 100-1000 micromolar applications stimulated ethylene production, respiration rates and ripening. Ripening can also be stimulated by picking the green, fully grown fruits.

Shortly afterwards,most fleshy fruits are abscised, unless they have already been taken by primates, birds, bats, etc. On the ground, they are subject to consumption, dispersal and decay through the activities of many other forest organisms. Once again, only a relatively small proportion of the total survive, and sometimes the entire fruit crop may fail to produce a single viable seed. The same can occur if seed insects or fungi are widespread, as frequently occurs in *Triplochiton scleroxylon*, where the fruit is non-fleshy and single-seeded. Where such fruits contain many seeds, the ripening process involves drying, and abscission layers in the fruit wall generally allow the seeds be shed while the fruit is still attached to the tree. In conifers, the ripe female cones open their scales as they dry.

The period from anthesis to fruit ripening has been studied in West Malaysia from the literature and observations over a 2 year period. The average interval was 4.3 months for 86 indigenous and 7 exotic species, with *Pterocymbium javanicum* showing the shortest period. *Firmiana malayana* and *Dillenia suffruticosa* also ripened their fruits rapidly. Those taking 9 months or more to ripen included *Diospyros maingayi*, *Palaquium hispidum* and *Vatica ridleyana*. The New World species *Bertholletia excelsa*, *Enterolobium cyclocarpum* and *Pithecellobium saman* have also been reported to ripen slowly. In West Africa, *Lovoa klaineana* has a short fruiting season, and *Mimusops heckelii* a long one. *Bosqueia angolensis* and *Celtis* spp, may fruit twice a year, while *Trema guineensis* fruits continuously. Although there may be a good deal of within-species variation, data of this kind can help in planning fruit collections, based on the time since flowering, which is often relatively easy to detect. However, not every flowering event is followed by a fruit crop.

There are fruits growing and ripening all the year round in tropical forests, which is important for the survival of frugivorous bats, in particular, since they have to maintain their body temperature by regualr feedling. However, there are usually *fruiting seasons* when a higher proportion of the species and individuals is in fruit. In Ghana, for instance,this is clearly during the dry season, with a fairly steady build-up to March. Most seeds will therefore be dispersed before the start of the rains. In a forest at 850 m altitude in eastern Zaire, the 200 species showed least fruiting (8 percent) in May, at the end of the 9 month wet season, and most fruiting (46 percent) in February.

Two dispersal peaks in the year were found by Foster for the canopy trees at Barro Colorado Island, Panama, one in March/ June and one in September/October. Lianes ripened fruit once a year, primarily between March an May, while understory trees and shrubs produced most ripe fruit in November/December. In discussing the significance of these seasonal peaks. Foster recognizes shorter and longer fruit development periods associated with species that flower at different times, such that they may disperse their seeds together. Also noted is the effect of whether variation, greatly affecting the animals normallly depending on this food supply.

In West Malaysia, ripe fruits were most frequently seen in September and October, and were least common from January to May. Of the 6 related *Shorea* spp that flowered successively the slowest rate of fruit development was found in the first to flower, and *vice versa*, with the result that all species bore ripening fruit at about the same time. Besides being a remarkable example of precise physiological control over the timing of reproduction, this result suggests that there may well be selective advantage in dispersing seeds at a particular time of year, even where the climate is realtively uniform.

However, by no means every genus distributes its fruiting periods in the same fashion- as with flowering there are instances of long-sustained fruit ripening, and of regular or irregular dispersal at intervals from a few months to many years. In West Africa, for example, fruits of *Terminalia ivorensis* have been collected from different sites in most months of the year, but on the other hand the fruiting of *Claoxylon hexandrum* is so regular that it is used to mark the date of a festival.

Seed Germination and Dormancy

A number of different types of seed germination can be recognized, base both on structure and function. In a study of more than 200 West Malaysian tree species, representing about 8 percent of the indigenous tree flora, Ng(1978) found that the hypocotyl elongated in 72 percent, carrying the cotyledons well above the ground surface. Most of these were the normal *epigeal* type, with cotyledons exposed and green, as in *Shorea leprosula*, *Koompassia malaccensis* and *Podocarpus nerifolius*, and many species that colonzie clearings and margins. Two modifications were noted: in *Diospyros* spp. the cotyledons are abscised early, before turning green; and a *durian* category is recongized, for example, in *Durio zibethinus*, *Dipterocarpus oblongifolius* and *Strombosia javanica*, in which the cotyledons remain within the endosperm and are abscised later. They also stay inside the seed coat in the normal *hypogeal* type, in which the hypocotyl does not elongate significantly, as for instance in *Anisophyllea*, *Garcinia* and *Lithocarpusl.* Here again, a modification is reported, with trees such as *Eugenia grandis*, *Lansium domesticum* and *Pithecellobium* spp. showing *semi-hypogeal* germination, with the cotyledons exposed at ground level.

Germination can also be classified in other ways. In a few species, the embryo has already developed into a viviparous seedling before becoming detached from the parent plant : *Dryobalanops aromatica* and several mangroves are examples of such *prior* germination. A much larger number exhibit *prompt* germination, in which the radicle emerges within a few days or weeks of dispersal. This is characteristic of dipterocarps, for example, and many other tree species,particularly those having large seeds that contain considerable food reserves.

Another class of prompt germinators are those species whole seeds germinate more rapidly and to a higher final percentage after thay have passed through the gut of a herbivore. Such stimulation was found for instance in *Azadirchta indica*, *Nauclea latifolia*, and especially strongly in *Securinega virosa* seeds germinated from baboon dung in a dry forest/savanna woodland area in Ghana, compared with seeds taken from fruits that had not been ingested.

The simplest reason for *delayed* germination is that the seeds were formed in fruits that dried on repening, and since being shed have not yet come in contact with moist soil or rain.

All the other categories involve some type of dormancy, such that most or all of the seeds fail to germinate in moist conditions at normal ambient temperatures. This relatively simple idea, of seeds being 'dormant', is frequently confused in the literature with their being 'viable', or 'not germinating', concepts which are related but not synonymous. The testa, and sometimes the indehiscent pericarp, may be impermeable to water until weakened by the combined effects of repeated cycles of heating and cooling, mechanical abrasion by mouth-parts or soil particles, chemical attack by micro-organisms. Once water can penetrate, such seeds become imbibed and usually germinate at once, but the key ecological significance is often that the germination of a cohort of seeds is spread out over a period of months, years, or even decades, persumably increasing the chances of some of the progeny surviving. 'Hard' seeds are common in the Leguminosae, although they are not formed for instance by *Koompassia malaccensis.*

Scarification is often used to pre-treat such seeds before sowing, and this increased germination of *Ochroma lagopus* from 3.5 percent to 84 percent. Surprisingly, effective techniques also included placing the balsa in boiling water for up to a quarter of an hour,or subjecting them to dry heat between about 60°C and 120°C. It appears that this type of breaking of dormancy could be one of the reasons why balsa seedlings frequently spring up after the passage of 'light' fires. Other seeds are freely able to take up water, but need to remain for a time in a dry condition for after-ripening to take place. Oil-palm, and freshly harvested cereals, including rice, generally do not germinate readily, but will do so after being kept indoors for a period.

In the common annual grass *Dactyloctenium aegyptium*, dry storage at 20°C for 10 weeks partially relieved this type of dormancy, while 40°C was much more effective. Vazquez-Yanes and his colleagues have elegantly demonstrated and investigated light-sensitivity in colonizing tree species in Mexico. The seeds of *Cecropia obtusifolia* and *Piper auritum* can remian dormant for more than a year if the imbided seeds are kept in the dark. They respond to light by germinating, but unlike some of the classical responses of herbaceous species a substantial r iod of illumination is required.

Nevertheless, the usual red/far red system involving the pigment wavelength and its stimulatory influence can readily be counteracted by following it with far- red irradiation. Even if red light was given for several hours, the germination percentage of *C. obtusifolia* only reached about 25 percent, but values of around 90 percent were achieved when seeds of either species were given 2 h of red light on each of 4 successive days.

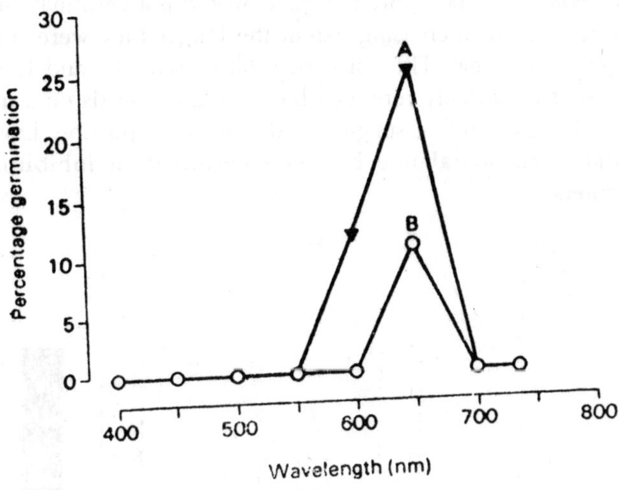

Fig. 6.8. Wavelength dependence of the breaking of seed dormancy by light, in (A) Cecropia obtusifolia; and (B) Piper auritum. Seeds were kept in the dark, except for 6 h irradiation at different wavelengths, and the final germination percentages were recorded after 1 month.

As discussed low red/far ratios probably act to keep such seeds dormant within the forest or in small gaps, but the higher ratios found away from the margins of larger clearings permit germination. The same group have demonstrated that a requirement for diurnal temperature fluctuation may be another mechanism involved in the breaking of seed dormancy of colonizer species in large gaps. For instance, imbided seeds of *Heliocarpus donnellsmithii* mostly remained dormant when kept under constant temperatures in the laboratory or buried in soil within the forest at about 25°C. High germination percentages were obtained under fluctuating laboratory temperatures, especially with a 6 h period at 32-39°C, and 18 h at levels 10°C cooler. Field germination was also enhanced near the middle of a sizeable gap, where the surface soil temperatures showed substantial diurnal fluctuations. Futher research can be expected to show whether other types of seed dormancy operate in tropical forest trees.

Combinations of more than one type of class tend to occur in the same seeds, and it is not unusual for dormancies to change during storage. Nor is it always easy to correlate the results obtained in the laboratory with conditions in the forest. For example, *Trema guineensis* seeds stored at 22°C would not germinate initially, but did so to an increasing extent the longer they were stored, at least up to one year. The dormancy still present at 6 and 10 months could be substantially removed by placing the seeds on a medium containing gibberellin, suggesting that it might possibly have been mediated *via* a balance between germination inhibitors and promoters.

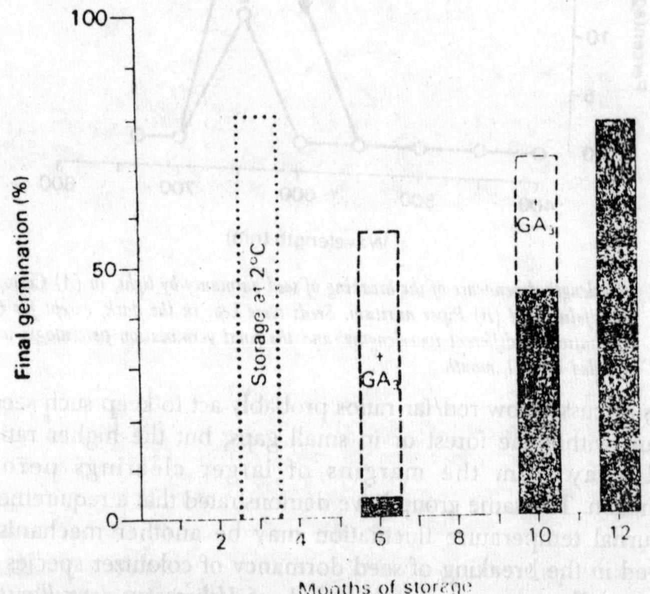

Fig. 6.9. Effects of storage time and temperature on the germination of Trema guineensis seeds, with and without gibberellin. Black column–storage at 22°C; dotted column–storage at 2°C; dashed column–promotion when seeds were germinated on agar plates containing 500 ppm gibberellic acid (GA$_3$).

Surprisingly, storage at 2°C markedly promoted germination, rather as winter chilling breaks the dormancy of tree seeds in north temperature zone forests. Additionally, there was an obligate requirement for light, with no germination whatsoever being obtained in the dark series. The germination temperature itself is also important, since all plants are thought to have minimum, optimum and maximum temepratures for this process, and germination rates are also strongly influenced temperature. For instance, germination of coffee took three months at 17°C, but only three weeks at 30°C.

Germination tests are generally carried out in convenient, standard conditions, for example at 25°C or 30°C, but as already mentioned these may not be appropriate for all species. The same applies to the question of suitable storage conditions: sometimes the dormant state may remain unchanged for considerable periods, as for instance in seeds requiring after ripening which may retain their dormancy if stored in a deepfreeze. More commonly, there is a continual modification of the degree of dormancy with the passage of time. Storage at 50 percent and 70 percent relative humidity at 22°C led to the development of a secondary type of dormancy in *Hildegardia barteri*, which was not the case when the fruits were stored at 90 percent relative humidity. This effect could be largely removed by placing the drier seeds into a moister atmosphere, and it apparently invloves changes in the permeability of the structures surrounding the seed itself. If may be that this mechanism operates under natural conditions on the stoop, rocky slopes colonised by this species.

During the dry season, the majority of seeds may perhaps remain unresponsive to occassional rain or dew, but in the moister conditions of the rainy season they may lose their dormancy, and germinate with a greater chance of survival. Storage temperatures and humidities are therefore the main considerations, together with avoidance of pests and diseases, when seeds are to be kept for use at a later date, particularly if long-term storage is planned for conserving gene resources.

An unusually extended longevity of nearly half a century has been reported for *Ochroma lagopus* seeds from herbarium sheets. However, in 'recalcitrant' species, which include many of the prompt germinator class, viability can be quickly lost if the seeds are dried, and they either germinate or go mouldy if kept at high humidity. Nevertheless, it now seems likely that more thorough research may indicate ways in which some of these species may be stored. For example, viability can be preserved for a month or more if dipterocarp fruits are disinfected and stored at about 95 percent relative humidity in closed containers. Temperatures must be above 15°C for *Shorea ovalis* and above 4°C for several other species, of which *S. talura* was the most tolerant. For oil palm seeds, formerly considered to be recalcitrant, the excised embryos can be thoroughly dried and kept at–196°C, where it appears that viability may be preserved indefinitely. *Agathis hunsteinii* seeds kept better at a moisture content above 32 percent, and die if dried to about 14 percent, but *A. cunninghamii* appears to be relatively easy to store. In *Triplochiton scleroxylon*, the speed of drying is an improtant consideration : when fruits were dried to 8 percent moisture content at a rate not exceeding 1 percent of the original weight lost per hours, they could be stored at–18°C for at least 18 months.

7

Growth Regulators

Among the substances which markedly influence the reactions and metabolism of plants are those internally synthesized compounds called *hormones.* In general, the term "hormone" is used to designate certain organic compounds which exert important regulatory effects upon the metabolism of an organism when present in only minute quantities. In the animal body it is generally considered that a further characteristic of a hormone is that it exerts its effects at a site remote from the locus of its synthesis. Adrenalin, for example, secreted in higher animals by the adrenal gland, has pronounced effects upon the heart and vascular system. Many, but not all, plant hormones similarly exert their effects in cells at some distance from those in which they are synthesized, after translocation from the latter to the former.

Included among the plant hormones are certain substances usually classified as vitamins in considerations of animal physiology. In the higher plants those vitamins which are essential exhibit essentially the same general type of behaviour as hormones and may be regarded as falling into the same general category of substances. In addition to the naturally occurring hormones in plants, a number of organic compounds are known which, when introduced into plants in relatively small quantities, induce effects which are similar to, and often apparently indentical with, those induced by naturally occurring hormones.

By an extension of the original concept, such substances are also often called plant hormones, although the more rigorous definition of the term would strict it to naturally occurring compounds. Other terms commonly used to designate plant hormones are *phyto-hormones*, *growth hormones*, *growth substances*, and *growth regulators* (a term which includes both *growth activators* and *growth inhibitors*).

Emergence of the Hormone Concept

Soding (1923), after Paal's deduction that specific substances produced in the coleoptile tip were responsible for phototropism, established that these very substances were capable of stimulating straight growth as well. Among the workers attracted by the demonstration of a correlation carrier in oat tips were Cholodny (1927) in Russia and Went (1928) at Utrecht, who independently extended the correlation carrier theory to both phototropism and geotropism. They concluded that all tropisms were mediated by a growth hormone system which was essential to all plant growth.

Went, while working on the role of auxin in plant growth discovered that hormone could be collected in an agar block by diffusion, and also presented a technique for obtaining the hormone from a great variety of plant materials. Using oat coleoptile he established a test so accurate and reproducible that it still stands as the best auxin assay technique, the *Avena* test. The sound basis which Went gave to auxin physiology resulted into a great deal of very productive work in subsequent years, and in a short eight-year period, from 1928 to 1936, three auxins were isolated characterized and identified, the quantitative relationships of auxin to tropisms of roots and shoots were established, and at the end of this period half of the significant and major functions of auxin in growth and development had also been discovered.

Kogl (1933) and his co-workers found two materials, strongly active in the *Avena* test, which they named auxin *a* and auxin *b*. Auxin *a* has a molecular weight of 328, which is very close to Went's figure for diffused auxin from *Avena*, i.e. 372. These compounds, today are only of historical interest since they have never been positively isolated from growing plant tissue. Kogl group (1934), as a result of further research, discovered another auxin compound which was identical to indole 3-acetic acid. The presence of this auxin was demonstrated in *Rhizopus* cultures by Thimann (1935). It is now generally accepted that indole acetic acid is the common growth hormone in higher plants.

Since the time of Went several physiological roles of auxins have been clarified. Went (1934) himself discovered that auxins stimulated the formation of adventitious roots (throwing light on another correlative effect). The developmental importance of auxin in morphological differentiation came to a full realization when Skoog and Tsui (1948) found that relative auxin levels in plant tissues play a crucial role in determining growth pattern.

La Rue (1936) found that auxins applied to leaves could retard leaf abscission. Later researches have proved that abscission of all plant organs (leaves, flowers, fruits, etc.) is correlated with low auxin content. The development of knowledge of the auxins has had a remarkable effect on the agricultural sciences as well. A historical outline of the early and recent contributions to our knowledge of plant growth hormones has been given by Boysen-Jensen (1936).

Definitions and Consideration of Nomenclature

It will be convenient at this point to turn to a consideration of nomenclature, since a wide range of terms have been used to characterize plant growth regulating substances. It is generally agreed that an ideal system of nomenclature must be based on their biochemical and physiological attributes, irrespective of their mode and site of production and processes involved in transporting them to the places where they act. Such an ideal nomenclature must therefore embrace not only naturally occurring compounds, but also all artificially made substances having the same action in the cell. A recent treatise by van Overbeek (1950) has given a very broad definition of plant hormones as "organic compounds which regulate plant-physiological processes regardless of whether these compounds are naturally occurring and/or synthetic, stimulating and/or inhibitory, local activators or substances which act at a distance from the place where they are formed.

Similarly, the term auxin is made to include all synthetic compounds having the specific auxin action. Before entering into an extensive discussion of these compounds, it will be well to define these terms. Out of the recent definitions of a plant hormone is given by Thimann (Pincus and Thimann, 1948) as "An organic substance produced naturally in higher plants, controlling growth or other physiological functions at a site remote from its place of production and active in minute amounts "Thimann favours the term "Phytohormone" (Greek : Phyton = a plant) which literally means a plant hormone. *Growth hormone* is the phytohormone involved in growth. The term auxin was originally suggested to refer to substances which were capable of promoting growth in the manner of the growth hormone. The term *growth regulator* refers to organic compounds other than nutrients, small amounts of which are capable of modifying growth. Thimann Pincus and Thimann, (1948) has defined auxin as "an organic substance which promotes growth (*i.e.*, irreversible increase in volume) along the longitudinal axis when applied in low concentrations to shoots of plants freed as far as practicable from their own inherent growth-promoting substances. Auxins may, and generally do , have other properties, but this one is critical." This definition is widely accepted among plant physiologists today.

The Extraction, Detection, and Estimation by Bioassay of Auxins

Although a detailed account of extraction and estimation techniques of auxin is beyond the scope of this text, a brief account may however be given. All procedures for the determination of auxin content of plant materials do not measure the same constituents. There auxin concepts are, however, not entirely simple. There are many examples were bound auxin is released in free form during extraction. Consequently it appears that the free and bound forms are in a dynamic stage and their strictly separate measurement is often difficult. There are two well defined methods for the extraction of auxins from material.

Diffusion method

Diffusion method, the simplest of all, consists of obtaining the growth hormone from plant material by diffusion into agar. The procedure consists of severing the growing tip or other organ to be tested under conditions which discourage transpiration and placing the cut surface for a period of an hour or so on a block of agar, usually of 1.5 per cent concentration. Three major difficulties may arise in the operation of diffusion techniques :

(a) The excessive loss of water, or a negative tension in the vascular system can prevent the accumulation of the diffusate in the agar block.

(b) The destruction of the auxin at the cut surface frequently interferes with the quantitative yield. Such destruction is enzymatic and is attributed in some cases, to polyphenol oxidase and in other cases to peroxidase.

(c) Growth inhibitors, whose existence is common in many green plants may prevent the effective use of diffusion techniques.

Solvent extraction method

Several solvents have been employed for the extraction of auxins from plant material. Among the solvent used are, chloroform, diethylether, ethyl alcohol and even water. A serious drawback to the use of chloroform, however, is the slow accumulation of a toxic substance, perhaps an auxin in activator, thought to be chlorine Thimann and Skoog (1940). Boysen-Jensen(1936) demonstrated that diethylether may serve as the most satisfactory solvent. But before use the ether should be redistilled over ferrous sulphate and calcium oxide in a small amount of water. This is done in order to avoid an oxidation of easily oxidizable auxin by the presence of spontaneously formed peroxide. It has been amply demonstrated that the formation of new auxin during extraction may be a source of serious error. This can be successfully avoided by using Gustafson's (1941) technique which consists essentially of boiling the plant material a short time prior to solvent extraction. The main features of his technique are as under:

1. Freezing the tissue rapidly on carbon dioxide ice.
2. Grinding the tissue with mortar and pestle.
3. Dropping the tissue into boiling water and allowing one minute of active boiling.
4. Collecting the plant material on a filter paper and extracting with ether for 16 hours, using 3 changes of ether.

Gustafson's technique has some limitations, since as pointed out by Thimann, Skoog and Bayer (1942), certain amount of the free auxin must be destroyed during heating, and by Van Overbeek, *et al.* (1945) that heating may cause the release of inhibitors which interfere with auxin assay. The formation of new auxin during extraction is also avoided by the use of freezing and lyophilization technique first described by Wildman and Muir in 1949. Their technique may be summarized as follows:

1. Plunging the plant material into liquid air or dry ice and acetone for rapid freezing.
2. Drying by lyophilization.
3. Grinding to 40 mesh in Wiley mill.
4. Storing in vacuo over phosphorus penta-oxide, in darkness, until use.
5. Extracting with peroxide-free ether (95 percent) at 0°C for four half-hour intervals.

6. Combining ether extracts and reducing in volume to a few ml.
7. Transferring qualitatively to agar for *Avena* assay. Some plant materials may not require either boiling or lyophilization for solvent extraction of free auxin.

Van Overbbeek *et .al* (1945) have developed a technique for obtaining free auxin using short term extraction. It may be epitomized as below:

1. Freeze plant material on CO_2 ice.
2. Slice bulky tissues into 2-5 mm slices.
3. Extract with peroxide-free ether at 0°C for two half-hour intervals.
4. Combine the ether extracts and reduce volume by evaporation to a few ml.
5. Transfer quantitatively to agar for *Avena* assay.

All the above techniques have been designed for the extraction of free auxin. Bound auxins can be extracted by using either of the latter two methods by carrying on extraction over a period of time.

Relation of Auxins to Growth of the Oat Coleoptile

The *auxins* have been the most comprehensively investigated group of plant hormones. Their action was first clearly demonstrated in the leaf sheath or *coleoptile* of the oat plant (*Avena sativa*). This is a tubular, leaf-like structure, closed at the top, which is the first part of the plant to emerge from the soil. Similar coleoptiles develop during early seedling growth of other members of the grass family. The coleoptile encloses the initial leaf and is eventually

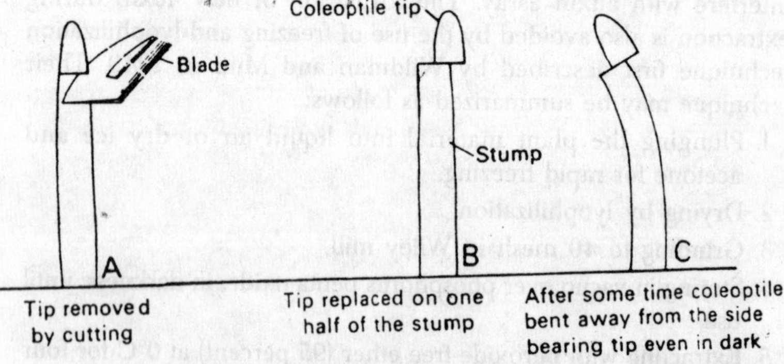

Fig. 7.1. Discovery of auxins. Experiment of Paal-eccentric placement of tip caused curvature.

pierced at the tip as a result of the growth of this leaf, soon after which all growth in length of the coleoptile ceases. Oat coleoptiles are approximately 1.5 mm. in diameter, and when illuminated seldom attain lengths of more than 2 cm. In the dark they may attain height ranging up to 6 cm. Cell divisions cease relatively early in the life history of an oat coleoptile, and during approximately the last three- fourths of its growth period all increase in its length results from cell elongation. If the tip of a coleoptile is removed by a clean cut made several millimeteres below the apex, the rate of elongation of the stump is immediately retarded.

If, however, the cut-off tip of the coleoptile or a similar tip from another coleoptile is affixed on the stump, its elongation will be resumed, and may nearly regain the original rate. Retipping the coleoptile with a short segment cut out of another coleoptile somewhat below the apex results in little or no increase in elongation rate. Such experiments indicate that the elongation of a coleoptile, which occurs in the more basal regions, is maintained only under the influence of some sort of a "stimulus": orginating in the tip, whence it is transmitted basipetally (apex to base) through the coleoptile. Went (1928, 1935) placed the cut-off tips of oat coleoptiles on a thin layer of 3 percent agar and after 1 hr, removed them and sliced the agar into a number of equal-sized small blocks. If

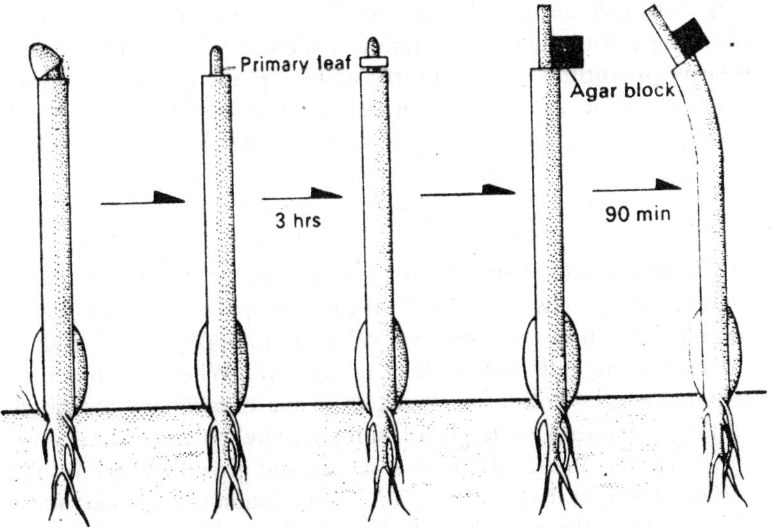

Fig. 7.2. Discovery of auxins. Experiments of F.W. Went–Avena Curvature test.

one of these blocks was placed upon the stump of a detipped coleoptile, the rate of elongation was accelerated just as if the stump had been capped with a fresh coleoptile tip. On the other hand, retipping a coleoptile with a block of pure agar had no appreciable accelerating effect on elongation.

It seems evident from these results that some substance or substances were transported out of the cut-off tip into the agar block, and subsequently out of the block into the coleoptile stump, whence they were translocated downward to the elongating region of the coleoptile. The substances which induce such a reaction now are classed as auxins. Many other plant organs such as stems, petioles, flower stalks, and coleoptiles of other species behave similarly upon removal of the apical region. Elongation is stopped or retarded by such a treatment but will be resumed if the excised apex of the organ is carefully relocated on the cut surface of the sump.

Biological Tests for Auxins

Auxins are known to be of widespread distribution in plants. They occur in such small quantities, however, that detection of their presence in an organic material by chemical methods is usually impossible. Recourse is had, therefore, to sensitive biological tests in order to demonstrate the presence of these substances. Several such tests have been used rather widely. The oat coleoptile test has been the most extensively employed method of determining the relative quantities of auxins present in plant tissues or other materials. If an agar block containing auxin from one source or another is affixed onesidedly on a detipped oat coleoptile elongation is found to be more rapid on the side of the coelontile below the portion of the tip on which the block is perched, resulting in curvature of the coleoptile.

Translocation of the hormone is almost strictly longitudinal, the elongating cells on the side of the coleoptile under the block receiving much more auxin than cells on the opposite side, with a corresponding differential effect on growth. When the block is centered on the coleoptile stump, as in the experiment described in the preceding section, all sides of the coleoptile receive approximately equal quantities of auxin, and growth proceeds in a vertical direction. Furthermore, it has been found that the curvature resulting from the eccentric attachment of agar blocks to detipped oat coleoptiles is proportional, within the range of about 0 to 20

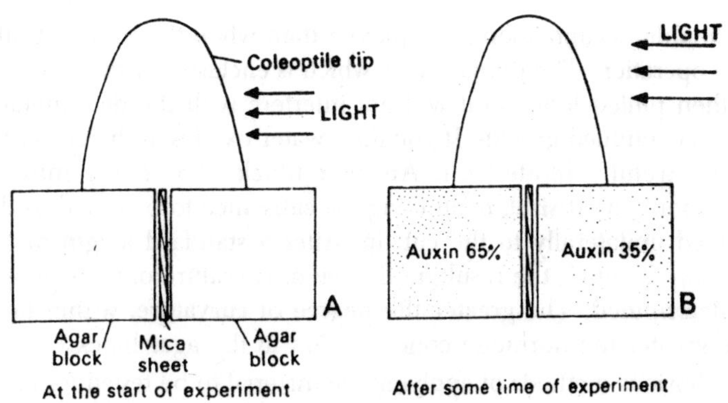

Fig. 7.3. Differential distribution of auxin in coleoptile tip under the influence of unilateral light.

degrees, to the concentration of the auxin in the agar block. It is this proportionality between hormone concentration and curvature that makes possible the use of oat coleoptiles as living test objects in the quantitative estimation of the auxin content of plant tissues or other materials.

Quantitative measurements of auxins by the oat coleoptile technique must be carried out under carefully standardized conditions. The oat seedlings (usually from a genetically uniform variety) are grown in a dark room at a temperature of 25°C, and a relative humidity of 90 percent. All manipulations are performed under phototropically inactive orange or red light. The Coleoptiles are used when about 2.3 to 4cm in length. The extreme tip of the coleoptile is first cut off, and after 3 hr, the topmost 4mm of the stump is removed. For reasons which cannot be considered in a brief discussion, the coleoptiles are more sensitive when this method

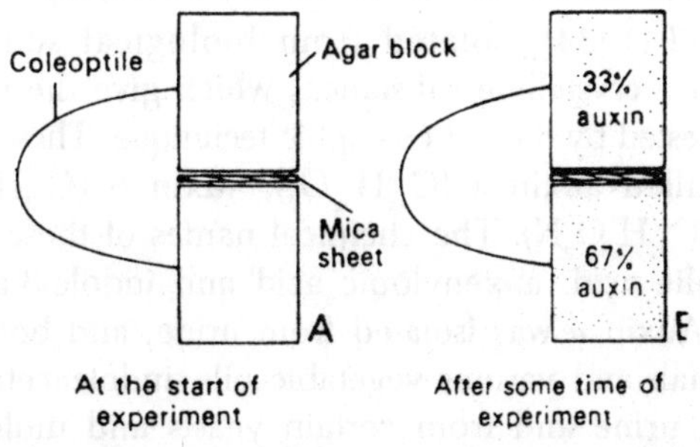

Fig. 7.4. Differential distribution of auxin from coleoptile under the influence of gravitational force.

of double decaptitation is employed than when the tip is cut off in one operation. The primary leaf, which is enclosed by the coleoptile, is then pulled loose so it will not interfere with the determination by its continued growth. If guttation water exudes at the cut surface it is carefully blotted off. An agar block (2 x 2 x 1 mm is a commonly used size), continuing the substance to be tested, is then affixed unilaterally to the cut tip. After a standard length of time (usually 90 min), the resulting degree of curvature of the coleoptile is determined. The greater the degree of curvature, within limits, the greater the hormone concentration in the agar block.

Various methods of applying the material to be tested for auxins to detipped coleoptiles have been employed. Sometimes small plant organs or pieces of plant organs are affixed directly to the cut surface of the coleoptile. More commonly plant tissues are placed in contact with 3 percent agar into which the hormone will move. The tissue is generally left in contact with the agar for about 2 hr. The agar is then cut into block of standard size each of which is then affixed unilaterally to a detipped coleoptile. The effects of pure chemicals or extracts can be tested by first dispersing them in agar, and them, after solidification, determining the influence of standard sized blocks of this agar on curvature of the coleoptiles. Or agar blocks can be soaked in a solution of the substances and then used in the coleoptile test for auxins.

Another commonly used biological test for auxins is to immerse sections of young oat coleoptiles in the solution to be tested and measure their increase in length over a certain period of time (commonly 6=25 hr.) as compared with increase in length of similar section immersed in water or an auxin free solution.

Chemical Constitution of the Naturally-occurring Auxins

Kogl *et al* (1934) isolated from biological sources three chemically pure crystalline substances which give the reactions of auxins when tested by the oat coleoptile technique. These substances have been called auxin a ($C_{18}H_{32}O_5$), auxin b ($C_{18}H_{30}O_4$), and heteroauxin ($C_{10}H_3O_2N$). The chemical names of these substances are auxentriolic acid, auxenylonic acid and indole-3-acetic acid, respectively. Auxin *a* was isolated from urine, and both auxins *a* and *b* from malt and various vegetable oils, indoleacetic acid was isolated from urine and from certain yeasts and molds. At one time it was considered possible that indoleacetic acid did not occur in the tissues of higher plants, but more recently it has been isolated from corn grains and there are numerous indications that it occurs in many other tissues of higher plants including oat coleoptiles. It seems probable that this will prove to be the principal naturally-occurring auxin.

The Occurrence and Synthesis of Naturally-occurring Auxins in Plants

Auxins appear to be universally present in plants, and their occurrence has actually been demonstrated in a wide variety of species. Furthermore, the auxins are nonspecific in their action, i.e., the same auxin, chemically speaking, which influences growth phenomena in one species usually also influences the same or similar phenomena in other species.

Auxins are present in plant cells in several different forms: free auxins, auxin precursors, and bound auxins. Various methods of extracting the "total" auxins (all forms) from plant tissues have been devised and more or less successfully employed. Other investigators have devised methods which they believe extract only the free auxins from plant tissues.

Only in the free form is an auxin readily diffusable, and only in this form is it effective in the oat coleoptile tests. Since, however, free auxins may be continuously formed from bound or precursor auxins, considerably more free auxin than originally present may diffuse into agar blocks from plant tissues. In most tissues thus far investigated the auxins present in inactive forms are many times greater than the free auxins present; the later seldom constituting more than 10 percent of the potentially available auxins.

The amino acid tryptophane has been shown to be a precursor of indoleacetic acid in plants. Conversion of the former compound into the latter apparently takes place through the intermediate stop of indoleacetaldehyde, which is therefore an even more immediate precursor of indoleacetic acid. The transformation of tryptophane to indoleacetic acid is catalyzed by a specific enzyme system which has been found in a number of plant tissues. The presence of zinc is necessary for tryptophane synthesis, and one of the indirect effects of zinc deficiency is a reduction in the quantity of auxin present. Following are the structural formulas for tryptophane and indoleacetic acids.

Many, but not all, of the bound auxins are auxin-protein complexes. Since, by suitable treatments, active auxins can be released from bound auxins, the latter are sometimes referred to as "auxin precursors," although they would not be so regarded in the usual sense of the term.

The auxins naturally present in plants are synthetic products of plant metabolism. The principal centers of auxin synthesis are apical meristematic tissues of aerial organs such as opening buds, young leaves, and flowers or inflorescences on growing flower stalks. Small quantities of auxins are also synthesized in apural root meristems although much of the auxin present in roots probably comes from aerial organs of the plants. The auxin synthesized in one tissue is frequently translocated to other organs of the plants. The concentration of auxin may vary greatly from one tissue to another; in general auxin is found in greatest concentration in the tissues in which it is synthesized or stored. Temperature is a factor in auxin synthesis, but the optimum for this process probably is not the same in all plants and tissues. In coleoptile tips the auxin is apparently synthesized from a precursor which is translocated atropetally (base to apex) through the coleoptile from the grain.

Auxins are not only synthesized in plant cells but are also inactivated in them. Inactivation may be brought about in various ways. Among other inactivation agents is a specific enzyme which breaks down indoleacetic acid; this enzyme has been isolated from the epicotyls of etiolated pea seedlings.

The Role of Auxins in Cell Elongation

Auxins play a role in the elongation phase of growth in many other plant organs similar to that described for the oat coleoptile. It is generally considered that cell elongation occurs only in the presence of auxins, and that with increase in auxin concentration there is an increase in the rate of elongation if no other factors are limiting. The optimum range of concentration for cell elongation varies greatly with different tissues, and relatively high concentrations usually exert an inhibiting, effect upon this phase of growth.

If the extreme tip of a maize or lupine root is cut off, its rate of elongation increases, although not greatly. Replacement of the root tip in maize plants results in a retandation in elongation rate at compared with detipped roots. Furthermore, attachment of coleoptile tips of maize to detipped root tips of the same plant results in a retardation in the elongation rate of the root tip. These results suggest that the same concentrations of auxin which accelerate elongation in coleoptiles and other aerial organs retard elongation in roots.

This supposition has been confirmed by experiments in which the roots of oat seedlings were immersed in pure solutions of auxins. The growth of the roots was found to be retarded in proportion to the concentration of auxin used. However, when roots which contain either no auxin at all or virtually none, are treated with auxin solutions of very low concentration, acceleration of growth as compared with similar but untreated roots often results.

The apparently contrasting effects of auxins upon elongation of roots and aerial organs may be explained by assuming that roots, buds, and stems all react in a comparable way to auxins their growth being inhibited by relatively high, and promoted by relatively low, auxin concentrations. Elongation of roots is favoured only at very low concentrations; at all higher concentrations their growth is checked. Stems and coleoptiles show a similar behaviour except that the optimum range of concentrations for elongation is much higher than for roots. The same concentrations of auxins which favour stem elongation result in retardation of root elongation. Briefly, therefore, whether an auxin will exert an accelerating or an inhibiting effect upon growth seems to depend in part upon its concentration and in part upon the specific tissue involved.

Shortly after the discovery of the naturally-occurring auxins various investigators showed that certain compounds, not known to occur naturally in plants, induce reactions in plants similar to those evoked by the naturally-occurring auxins. The list of such "synthetic" auxins has become quite extensive. The best known of these substances are α-naphthalene acetic acid, indolebutyric acid, 2,4-dichlorophenoxyacetic acid, and α-naphthoxyacetic acid. Most of these compounds induce curvature of oat coleoptiles when tested by the standard technique, although many of them are not as effective in causing this reaction as the naturally occurring auxins.

The term "auxin" has been used in different senses by different authorities. Most commonly, however, an auxin is considered to be any organic compound that promotes growth along a longitudinal axis when applied in low concentrations to organs which are initially low in their content of such growth-promoting substances but are under conditions which are otherwise favourable for elongation growth. The phrase "low concentrations" is a somewhat vague one, but in general can be taken to refer to concentrations of less than 10^{-3} molar. In spite of the apparent diversity of the compounds which act as auxins, certain similarities in molecular structure are common to all of them. These are : a ring system with a side chain containing at least one carbon atom between a terminal carboxyl or potential carboxyl group and the ring, a double bond in the ring adjacent to the side chain, and a definite space relationship between the carboxyl group and the ring system. *Cf.* the structural formula for indoleacetic acid given earlier.

Translocation of Auxins

If a block of agar containing auxin is affixed to the morphologically upper end of a segment of oat coleoptile, and a block of pure agar to the lower end, auxin will move into and accumulate in the lower block. The final concentration of auxin in the basally attached block may greatly exceed that in the one affixed to the apex. If the agar block containing auxin is affixed to the morphologically basal end, no translocation of auxin will occur. Translocation of auxin in oat coleoptiles apparently takes place through the paraenchyma tissues.

The results of such experiments show that transport of auxin in the oat coleoptile is *polar* i.e., occurs only basipetally, and that it can occur against a concentration gradient since the auxin

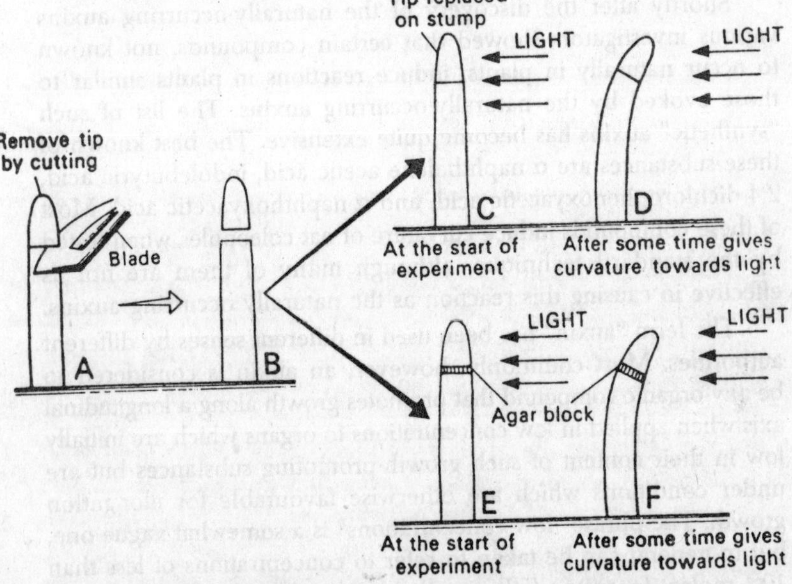

Fig. 7.5. Discovery of auxins. Experiments of Boysen-Jensen.

accumulates in the lower block. There is evidence that a similar basipetal translocation of auxin, either naturally-occurring or introduced, occurs in many other plant tissues or organs. Among these are coleoptiles of other species than oat, the veins and petioles of leaves, hypocotyls, herbaceous stems, and woody stems. In such organs the polarized downward movement occurs in parenchyma or phloem tissues. In roots on the other hand, movement of auxins appears to be nonpolar and there are undoubtedly other tissues in which this is also true. The mechanism of the polar movement of auxins is unknown. Etherization stops the transport of auxin, except insofar as it can be accounted for by diffusion, and destroys its polarity. This indicates that living cells are involved in the process, but tells nothing of the manner in which they operate.

It has been suggested that the polarity in the movement of auxins may result from differences of electrical potential in the tissues, but experiments designed to test this hypothesis have not yielded evidence in its support. Upward translocation of auxins can occur in plants, at least under certain conditions. If an auxin is applied to the soil or to the basal parts of entire plants or to cuttings in adequate concentrations, their absorption and upward movement through the plant can be demonstrated. Such upward transport apparently takes place only when the auxin molecules pass into the transpiration stream. As soon as the auxin molecules move back into the living tissues of the stem or leaf their polar basipetal movement is resumed.

Role of Auxins in Root Formation

It has been known for many years that the presence of buds on a cutting favours development of roots when the basal portion of the cutting is introduced into a suitable rooting medium. Developing buds are more effective in promoting root formation than quiescent buds. Leaves, especially if young, also often favour the production of roots on cuttings. These observations suggest that root initiation on cuttings is favoured by hormones which are synthesized in the buds and young leaves and are subsequently translocated to the basal part of the cutting. Soon after the identification of them as naturally occurring auxins it was found that auxin *b* and indolesacetic acid are active in inducing root formation.

There is good evidence that other hormones besides auxins are also necessary for root formation or atleast for their continued development, whether the roots are initiated on other roots, on stems, or on leaves. The effect of auxins on root *formation* should be clearly distinguished from their effect on root *elongation.* In general, the concentrations required for the former process are much greater than for the latter. A number of the "auxins" not known to occur naturally in plants have also been found to be effective in promoting root formation in many species.

Extensive experiments have been carried out on the suitability of treatments with various auxins as a practical method of aiding in the rooting of cuttings. Such treatments are not effective with all kinds of plants, but with cuttings of many species they lead to a speeding up of the process of root formation and to the development of a greater number of roots per cutting. Hormones do not induce root formation, however, on cuttings of species on which at least some roots do not develop without their application. Various techniques are used to introduce hormones into cuttings. Formerly the most favoured procedure was to immense the basal end of the cuttings in a dilute (10-200 p.p.m) solution of the hormone for periods ranging up to 24 hr, before setting them in the rooting medium. Laterally this method has been largely superseded by two less time-consuming procedures. In one of these the hormone is applied in a day form, being first mixed with an inert powder such as tale, most commonly in proportion of 500-2000 parts of the hormones to 1,000,000 parts of tale.

The basal end of the cutting is first dipped in water, and then in the powder before insertion into the rooting medium. In the "quick dip" method the basal ends of the cutting are dipped momentarily (for about 5 sec) into a relatively concentrated solution of the hormone (4,000-10,000 p.p.m, in water or 50 percent ethyl alcohol) before being a set in the cutting bench. Auxins many also be applied to stems or cuttings in lanolin or an vapours. The compounds most commonly employed in all of these methods, either singly or in mixtures, are α-napthaleneacetic acid, napthalene acetamide, and indolebutyric acid. The most effective treatment varies according to species; extensive tabulations of the effects of various compounds and methods of treatment on numerous species are given by Avery and Johnson (1947) and Mitchell and Marth (1947).

Other Effects of Auxins

One of the most remarkable facts about the auxins is the multiplicity of the growth reactions in which they participate. Only

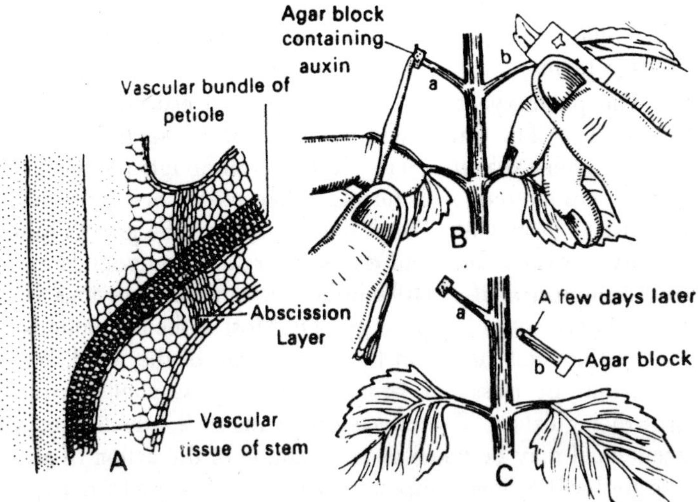

Fig. 7.6. Effect of auxins on abscission. A–A normal abscission layer at the base of petiole. B–(a) agar block contains auxin is fixed on debladed petiole (b) agar block without auxin is being fixed on debladed petiole. C–(a) There is retardation of abscission of petiole (b) Petiole abscised.

a few of their roles are discussed in this chapter; other important growth phenomena in which they play a part include the development of fruits, abcission of leaves and fruits, activation of cambial cells, apical dominance, flower initiation, geotropism and phototropism. These various roles of the auxins will be discussed in subsequent chapters.

Toxic Effects of Auxins

Reference has already been made to the inhibitory effect of relatively high concentrations of auxins on the process of cell elongation. Applications of auxins in relatively high concentrations also result in various kinds of growth malformations in plants such as distortions of leaves, stems and roots, discolourations of leaves, inhibitions of stem or root elongation or flower opening, and the formation of tumors. It should be emphasized that the term "relatively high concentration" as used in this discussion refers to concentrations which, in absolute terms, are very low, of the general order of magnitude of 1000 p.p.m. In somewhat higher concentrations, but in absolute terms still very low, these compounds often result in death of the plant. Realization that auxins, when applied in relatively high but actually very low concentrations, exert toxic or lethal effects on plants led to the suggestion that they could be employed for the purpose of killing weeds or other noxious plants.

The phenoxyacetic acids have proved especially effective as herbicides. The most widely used of these has been 2,4 - dichlorophenoxyacetic acid ("2,4-D"), although numerous other organic compounds with hormone-like effects or plants also show promise as herbicides. The results of tests on the growth-inhibiting activity of over a thousand different organic compounds are listed by Thompson *et al.* (1946),who also summarize previous work on the chemical control of plant growth. Most of the compounds which they list are not auxins in the usual sense of the word, but all of them may be classified, broadly speaking, as plant hormones or growth regulators. The effects on plants of 2,4-dicholorophenoxy acetic acid may be taken as an example of the action of an auxin-like herbicide. This compound is readily absorbed from sprays or dusts when applied to leaves.

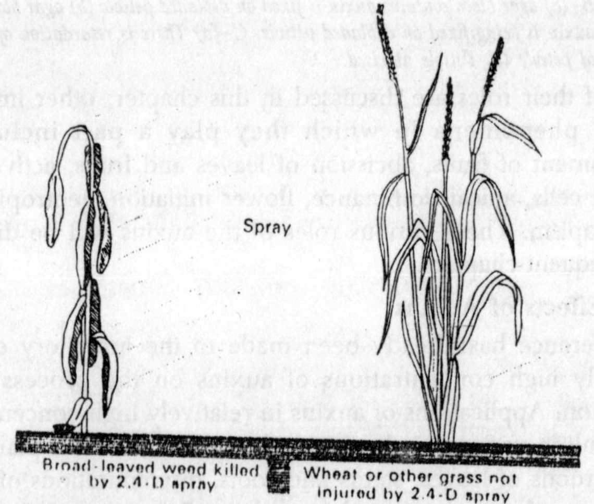

Fig. 7.7. Destruction of weeds by 2, 4-D.

It is quickly translocated to other parts of the plant and affects especially the meristems. The rapid distribution of this compound throughout the plant contributes greatly to its effectiveness as a toxicant. Death results from demagements of metabolism, especially in the meristems. Different kinds of plants differ markedly in their reactions to applications of 2,4 -dichlophenoxyacetic acid. Cereal

Fig. 7.8. Auxin spray on a plant prevents premature fruit abscission and increases the yield. A–Unsprayed, B–Sprayed.

grains and most other grasses are less susceptible, for exmple, than most broad-leaved annuals, and most woody plants are less susceptible than most herbaceous species. This very selectivity of effect is one of the advantages of this compound, and of many similar ones, as herbicides. Broad-leaved weeds can be elinimated from sugar cane fields or from lawns, for example, by application of a spray in the proper concentration and at the proper volume per square foot. Besides its selectivity of effect, 2,4-dichlorophenoxyacetic acid has certain other advantages when used as herbicide. The absolute concentrations required are very low.

Many plants can be killed, for example, by spraying with an 0.1 percent solution at the rate of 5 gal.per 100 ft^2. In the concentrations generally employed this compound is harmless to men and animals. Under most conditions, 2,4-dichlorophenoxyacetic acid disintegrates rapidly in soils. The possibility of detrimental effects on plants which subsequently grow on soils in which this compound was once present thus disappears rapidly. Great care must be exercised, however, in the use of this herbicide, that the spray does not drift to plants other than those being treated. Accidental damage or destruction of valuable plants as a result of careless application has been the chief disadvantage in the use of 2,4-dichlorophenoxyacetic acid as an herbicide.

The differential effects on various species of plants of some of the other compounds used as herbicides is unlike that of 2,4-dichlorophenoxyacetic acid. Isopropylpheny learbamate, for example, in contrast with 2,4-dichlorophenoxyacetic acid is much more effective in killing certain grasses than broad-leaved annuals

It is convenient to mention several other growth inhibitors at this point, although whether or not they would be classed as hormones is a matter of definition. Coumarin favours cell enlargement in such plant tissue as oat coleoptiles and leaf blades at very low concentrations, but checks cell enlargement at all higher concentrations. Similar checking effects upon root elongation have been found for certian coumarin derivatives. Another growth inhibitor which has attracted considerable attention is maleic hydrazide. Applied to plants in relatively low concentrations (about 0.4 percent) this compound results in general cessation of growth. At somewhat lower concentrations loss of apical dominance is evident, and axillary buds start to grow soon after treatment. At certain concentrations flowering appears to be largely suppressed without much suppression of vegetative development. Plants of various species seem to be affected in essentially the same way by this compound.

Mechanism of Auxin Action

The earlier attempts to elucidate the mechanism of auxin section mostly centered around proposed explanations of the part played by such compounds in cell elongation. Auxins appear to have two main effects in this process: they cause an increase in the plasticity of the wall and they participate, directly and indirectly, in the reactions whereby additional cellulose molecules are deposited within the wall. With the growing relization, however, of the large number of growth reactions conditioned by auxins, it has become obvious that these compounds must play a role in some pivotal metabolic process. The apparent necessity of specific chemical structure for a compound to act as an auxin, mentioned earlier, also suggests a particular metabolic role. The effect of auxins on cell wall development is, therefore, now generally considered to be, not a direct one, but only one possible end expression of fundamental auxin-conditioned or regulated metabolic processes.

There is considerable evidence that the auxins act primarily in a catalytic or regulatory capacity in some phase of the carbohydrate metabolism of plants. A suggestive finding in this connection is that introduction of auxins into leaves or cuttings induces marked hydrolysis of starch. Auxin also appears to participate in some part of the respiratory process but the exact nature of this relationship has not yet been traced. A significant fact in this connection is that auxins exert their effects only under aerobic conditions. The discovery by Wildman and Bonner (1947), that certain auxin protein complexes from spinach leaves possess phosphatase activity, may prove to be a significant one in allocating a specific role to the auxins. Although the question of the metabolic role of the naturally-occurring auxins as yet cludes a definite answer, it seems practically certain that they operate as carriers, coenzymes, or prosthetic groups in come fundamental enzyme system which plays a part in carbohydrate or organic acid metabolism.

The inhibiting as well as the accelerating effects of the auxins must be accounted for in any hypothesis of the mechanism of their action. If we accept the likely assumption that an auxin is active when it operates as the prosthetic group of an enzyme, it is reasonable to postulate that two fundamental properties of its molecules would be: (1) a specific group which reacts with the substrate molecules, and (2) the necessary structural configuration to combine in some manner with the protein portion of the enzyme.

Viewed in the light to this hypothesis (which incidentally is also applicable to other enzyme systems), relatively high concentrations of an active auxin may result in inhibitory effects as a result of some of the auxin molecules preempting all free positions on the protein component of the enzyme, while others become chemically attached to substrate molecules. This effectively blocks the reaction which can occur only when the protein and auxin operate as a catalytic team, so that substrate molecules become linked temporarily to the protein through auxin bridges

Inhibition of auxin-controlled reactions may also result from the presence in the metabolic system of growth-inhibitors with an auxin-like structure which possess strongly only one of the two properties listed above and the other one only weakly or not at all. If such a compound possesses only the capacity of combining with the protein, it may act as an inhibitor by occupying positions in the protein complex which would otherwise be taken by more active auxins. On the other hand, if the auxin like compound possesses only the property of reacting with the substrate, it may block the overall reaction which can only take place if the compound can act as a chemical bridge between the protein and substrate molecules.

Certain compounds, often called *antiauxins*, are known which offset or counteract the usual growth-enhancing effect of auxin in plants. Among these are coumarin and maleic hydrazide both mentioned earlier in this chapter as growth inhibitors. Another antiauxin is *trans* -cinnamic acid. The effect to this compound is of especial interest in view of the fact that its isomer *cis* -cinnamic acid has the properties of an auxin.

Other Plant Hormones

The auxins are only one kind of a number of hormones occurring in plants. Some of the other compounds in this category have been isolated from plant tissues ; the existence of others is inferred from the occurrence of various physiological reactions; other such substances doubtless are as yet undiscovered or unsuspected.

Traumatic acid

Haberlandt (1921) showed that if freshly cut plant tissue is immediately rinsed with water, very few cell divisions occur in the cells adjacent to the would. However, if the wounded area is smeared with finely ground tissue of the same species cell division takes place. This result led him to postulate the existence of substances which he called "wound hormones" in injured tissues, which are required if cell division is to be engendered in the cells bordering a wound.

More recently English et al. (1939) have succeeded in extracting from bean pods and purifying a compound called traumatic acid which acts like a "wound hormone", This compound has the following formula:

$$HOOC.CH = CH.(CH_2)_8.COOH$$

Traumatic acid induces extensive wound periderm formation in washed disks of potato tuber. It is probable that there are also other "wound hormones" present in plants.

Calines

Went (1938) has postulated the existence in plants of a group of hormones which he calls the *calines* : (1) rhizocaline, made in the leaves and necessary for root formation; (2) canlocaline, synthesized in roots, but necessary for elongation of stems; and (3) phyllocaline, made or at least stored in cotyledons, and necessary for leaf growth. It is also considered that rhizocaline and caulocaline may be stored in seeds. Most of the evidence for the existence of such hormones in plants is indirect and the effects attributed to these hypothetical hormones may actually turn out to result from the activity of substances already well known to have important influences on the growth or metabolism of plants. Glaston and Hand (1949), for example, have obtained evidence that adenine has effects on plants analogous to those of the postulated calines. The calines are discussed further in the following chapter.

Hormones of reproduction

Several hormones are known or believed to play a role in the reproductive process of plants. Among these are the ubiquitous auxins, the "embryo growth factor" and the postulated "florigen".

Vitamins

From the standpoint of human physiology the vitamins constitute a group of specific organic substances which must be supplied in diet, but which are required in relatively small, often only minute, quantities compared with carbohydrates, fats, and proteins. From the standpoint of living organisms in general, for reasons for discussed later, vitamins must be regarded simply as one rather arbitrarily delimited group of growth and metabolism regulators which cannot be sharply distinguished from the hormones. The principal substances recognized as vitamins in human and animal nutrition constitute a very heterogeneous group of compounds.

All of the vitamins, or at least their immediate precursors, are sythesized in green plants. The human body, on the other hand is dependent on sources outside its own metabolism for most essential vitamins and obtains a large proportion of them in green plants used as food. One partial exception is vitamin D which can be synthesized in the human body under irradiation from appropriate precursor which come from foods. Absence of deficiency of any one of the necessary vitamins in the human body results in physiological malfunctioning often evidenced as a specific disease. Some of the vitamin-deficiency diseases of man, such as scurvy, resulting from ascorbic acid deficiency, and beri-beri, resulting from thiamine deficiency, have been known for centuries, although only a comparatively recent years have their causes been recognized. Other higher animals appear to have vitamin requirements very similar to those of man.

The latest addition to the group of B vitamins, is vitamin B_{12}. This vitamin is of especial interest because cobalt is a constituent of the molecule. Whether of not this vitamin plays any role in plant metabolism is not known. Vitamin requirements differ from one kind of living organism to another, but no plant or animal is known for which at least some of them are not essential. The first fundamental question to be answered regarding the vitamin physiology of any organism is: Does the organism require this specific vitamin in its metabolism? Certain of the vitamins appear to be necessary in the metabolism of all plants and animals. This is true of many, if not all, of the vitamins of the so-called B complex. On the other hand, certain vitamins seem to be essential in the metabolism of only certain kinds of organism.

Table 7.1. The Vitamins.

"Alphabetical" Designation	*Name*	*Molecular Formula*
Vitamin A	Hydrolytic product of carotene	$C_{20}H_{29}OH$
Vitamin B (Complex)	Thiamine chloride hydrochloride (B_1)	$C_{12}H_{18}N_4SOCl_2$
	Riboflavin (B_{12} Vitamin G)	$C_{17}H_{20}N_4O_6$
	Nicotinic acid (niacin)	$C_5H_4N\text{-}COOH$
	Pyridoxine(B_6)	$C_8H_{11}NO_3$
	Pantothenic acid	$C_9H_{17}NO_5$
	Inositol	$C_6H_{12}NO_6$
	Biotin(vitamin H)	$C_{10}H_{16}O_2N_2S$
	Folic acid	$C_{19}H_{19}N_7O_6$
	p-amino benzoic acid	$NH_2C_6H_4COOH$
Vitamin C	Ascorbic acid	$C_6H_8O_6$
Vitamin D	Derivatives of ergosterol, cholesterol and other sterols	$C_{28}H_{43}OH$ (activated dehydrocholesterol)
Vitamin E	α-tocopherol (and others)	$C_{29}H_{50}O_2$
Vitamin K	K_1 : 2-methyl, 3 phytyl, 1,4-naphthoquinone	$C_{31}H_{46}O_2$
	K_2 : 2-methyl, 3-difarnesyl, 1,4-naphthoquinone.	

A second fundamental question to be answered regarding any essential vitamin for a given organism is: Does the organism synthesize this vitamin in quantities adequate to meet its own metabolic requirements? The answer to this question for the human organism is essentially "no" for all of the corresponding question for green plants is "yes". For bacteria and fungi no categorical answer is possible since the vitamin- synthesizing capacities of such organisms vary greatly from one species to another, and even from one variety of a given species to another.

For many years it has been known that, in the culturing of bacteria or fungi on artificial media, it is necessary to introduce some organic material such as potatoes, peptone, oat or corn meal, yeast extract, dung, or wood into the medium in addition to sugar and other pure chemicals if growth of the organism is to occur. Only in comparatively recent years has it been recognized that these organic materials are sources of essential growth substances, some of which, at least, are identical with the known vitamins.

Direct experimentation has demonstrated that many bacteria and fungi cannot synthesize certain of the vitamins which they require, or at least cannot synthesize them in sufficient quantities to permit optimum growth even when all other growth conditions are favourable. If the substrate on which they are growing is deficient or lacking in one or more of the necessary vitamins, development of the organism will be retarded and may cease entirely. By systematic addition of the various vitamins to a vitamin-free medium on which an organism is cultured it is possible to ascertain which of the essential vitamins that organism must obtain from its substrate.

It is probable that the vitamin requirements of all fungi are very similar; but different species differ greatly in their vitamin-synthesizing capacity. Many species, for example, cannot synthesize thiamine in adequate quantities. Hence, unless this compound is present in the substrate, the fungus will be retarded in development as a result of the thiamine deficiency. In capacity to synthesize adequate quantities of biotin, pyridoxine, pantothenic acid, have also been demonstrated for many species of the fungi.

It has already been mentioned that all of the known vitamins, or at least their immediate precursors, are synthesized in higher green plants. Whether or not all such compounds are essential in the metabolism of green plants is an open question. The evidence regarding this point must of necessity be somewhat indirect. There are no good reasons, however, for believing this to be true of vitamins A, D, E and K. On the other hand, thiamine, nicotinic acid, riboflavin and pyridoxine are known to be constituents of the prosthetic groups or coenzymes of important enzymes found in vascular green plants and are probably necessary in the metabolism of all living cells in such organisms. It is highly probable that some, and perhaps all, of the other B vitamins are required by green plants and ascorbic acid may also be an essential green plant vitamin.

Although the higher green plants appear to be self-sufficing with regard to necessary vitamins, and such compounds are widely distributed through the organs of such plants, this does not necessarily mean that all cells of all tissues synthesize these compounds in adequate quantities. Some parts of a plant are dependent upon other parts for certain necessary vitamins. Experiments on excised roots of certain plants growing in sterile culture show, for example, that thiamine must be added to the culture medium if normal growth is to continue. Obviously roots, at least of some species, do not synthesize enough thiamine to meet their own needs, and irract plants are presumably dependent upon down-ward translocation of this substance from leaves in which it is synthesized. Such downward translocation of thiamine, and also of pantothenic acid and pyridoxine, has been demonstrated in the tomato. This is probably also true of nicotinic acid. Upward translocation of thiamine from older to younger leaves also occurs. There is also evidence for a similar downward translocation of thiamine from older to younger leaves also occurs. There is also evidence for a similar downward translocation of ascorbic acid in plants. On the other had, roots of at least some species appear to synthesize certain other vitamins in such quantities as not to be dependent upon the tops for a supply of these compounds.

The growing embryo and endosperm in the developing seed of at least some kinds of plants appear, like roots, to be dependent upon translocation from other parts of the plant for necessary thiamine. In wheat, for example, as the content of this hormone in the developing grains increases, there is a corresponding diminution in the thiamine content of vegetative parts. It is possible that young developing embryos of some species may similarly be dependent upon vegetative tissues for other essential vitamins.

From the foregoing discussion it should be evident that no sharp distinction can be drawn between “vitamins” and “hormones” in considerations of the metabolism of higher plants. Some of the so-called hormones, such as traumatic acid, operate in or close to the cells in which they are synthesized;' others, such as the auxins, are frequently translocated to other, often distant cells, in which they influence metabolic processes. Similarly, some of the so-called vitamins, such as riboflavin, appear to operate principally in cells in which they are synthesized, whereas others, such as thiamine, are often present in some tissues largely or entirely as a result of translocation from other organs of the plant. From the standpoint of plan metabolism, all such compounds fall into the one comprehensive category of "growth-regulating substances" or "hormones".

The Effect of External Factors on Auxin Content

It is difficult to make generalizations of aid to plant physiologists owing to sparse literature on this aspect of auxin physiology. Nevertheless several such external factors may be listed, which influence the auxin content of plants:

Temperature

Few, if any, effects of temperature are interpretable in terms of auxins. Several workers in the past have attempted to study the relation between auxin content of the plant and temperature prevailing at the time of experimentation, but only a few of the recent ones are mentioned. Burton (1956) found no consistent relationship between storage temperature and ether - extractable auxin content in potato tubers. Blommaert (1955) and Hendershott and Bailey (1955) reported a decrease or total absence of auxin in winter-dormant buds of peach and marked resurgence in auxin content immediately prior to and during the onset of growth in the spring.

Mineral Nutrition

The relation between auxin content of intact plants and mineral nutrition has been studied in detail, only for nitrogen and zinc; deficiencies of either of the element lead to a lowered auxin content. The effect of nitrogen is considered to be an indirect one, since its effects appear after the appearance of visible deficiency symptoms or a decline in growth. Zinc probably affects auxin content possibly by reducing tryptophan synthesis.

Like nitrogen, effects of copper and manganese on auxin content of the plant appears to be indirect, because in copper and manganese deficient plants decreased auxin content is found only after the appearance of visible deficiency of these elements. Eaton (1940) suggested that boron deficiency reduced auxin level in cotton plants but MacVicar and Tottingham (1947) failed to confirm these results with cotton, tomato, sunflower, soybean and tobacco plants. The effect of other mineral elements on the endogenous auxin level is little studied and the facts thus, are incoherent.

Hydrogen-ion Concentration

Rufelt and Fransson (1956) studied the effect of pH on the auxin content of wheat roots, with solutions at pH 5.0-7.5. They noticed a short-term increase in auxin content, with a maximum at about six hours after transfer, and a subsequent restoration of the original level after 24 hours in the new solution. The mechanism of this effect is, however, obscure.

Ionizing Radiations and Ultraviolet

The publications on this aspect clearly indicate that X-irradiation can markedly and rapidly lower auxin content probably by interfering with auxin synthesis.

Skoog (1935) reported that moderate X-ray doses immediately reduced the amount of diffusible auxin obtainable from apices of *Pisum*, *Vicia* and *Avena* plants. Gordon's work (1956, 1957) has confirmed Skoog's observations of immediately decreased auxin content following X-irradiation but stresses the significance of reduced auxin synthesis. Spencer and Cabanillas (1956) reported that exposure of *Indigofera endecaphylla* seeds to either X-rays (80,000r) or thermal neutrons, both significantly reduce the auxin content of plants emerging from the treated seeds. The effect of ultraviolet irradiation has also been found to decrease endogenous auxin level.

Visible Light

A considerable work on the effects of visible radiations on the auxin content of plants had been undertaken. High light intensity, length of photoperiod and effects on etiolated plants, all have been investigated in detail. The photoinactivation of auxins has already been discussed in this chapter. It may be concluded that visible light has the effect of inactivating the auxin present in the plants.

Chemicals

Certain externally applied chemicals, such as ethylene, maleic hydrazide, and 2, 4-dichlorophenoxyacetic acid have been found to exet influence on the auxin content of intact plants. Lockhart and Weintraub (1957) showed that 2,4-D decreases the free auxin content of terminal shoot of *Phaseolus* seedlings. It is unanimously but unusually concluded that growth inhibitory properties of maleic hydrazided are not due to any effect on auxin content. Exposure of *Avena* coleoptile tips to ethylene has been found to decrease the diffusible auxin content by van der Laan (1934).

Gibberellic Acid

Nitsch (1957) reported that application of gibberellic acid to certain woody plants results in an increase in extractable auxin, Galston's observation that feeding of gibberellic acid (GA) to pea plants raises the level of an auxin inhibitor of the enzyme IAA oxidase leads us to believe that the overall effect of GA is that it has an auxin sparing action. Similar view was expressed by Stutz and Watenabe (1957) with *Lupinus albus.* Both gibberellic acid and long photoperiods were found to depress overall initial IAA oxidase activity, thus probably leading to a "sparing" of auxin by gibberellic acid.

Miscellaneous

Several other factors, in addition to those discussed above may be responsible for altering both the level and metabolism of the endogenous auxin in the plant. Among these may be mentioned: (i) abnormal growth and pathology (ii) root-nodule bacteria, (iii) crown gall, (iv) virus diseases, and (v) other pathogens. All these, in some way or the other, are known to alter auxin level of intact plants and plant parts.

Theories of the Mechanism of Auxin Action

It would appear to be over-simplication if we ascribe the full responsibility for growth to any one function in the plant. This is so because the action of auxins in the control of growth is a complex of many functions. Although the mechanism of these functions it may be concluded that none of them known till today can entirely account for the effects of auxin on growth.

The numerous extremely interesting physiological and biological responses to plant hormones, observed in the last 25 years account for the intricate bearing on their mechanism. Gordon (1952) enumerated some of the many responses which may be briefly summarized.

1. increase in plasticity of the shoot and elasticity of root cell walls,
2. increased permeability of the cell to water,

3. enhanced capacity to retain water taken up,
4. "active"; uptake of water and solutes,
5. decreased photoplasmic viscosity,
6. increased respiratory rate,
7. more rapid synthesis of proteins with lowered levels of free amino acids in the free amino acid pool,
8. increased accumulation of monosaccharides at the expense of polysaccharides, and
9. a stimulation of the activity of many enzymes, such as alcohol and malic dehydrogenase, catalase, phosphatase and ascorbic acid oxidase, and also several other responses.

Several theories which have been put forward to explain the mechanism of auxin action consider one of the above effects as the basic response on which other phenomena rest. For the sake of convenience, discussions of the hypotheses concerning the mechanism of auxin action will be grouped into five heads.

Molecular Reaction Theories

A molecular reaction into which auxin might enter in causing growth was first suggested by Skoog et.al (1942) who postulated that auxins may act as coenzymes, serving as a point of attachment for some substrate onto an enzyme controlling growth. A different suggestion for the role of auxins was forwarded by Veldstra (1955). He opined that the degree of fat solubility as influenced by the ring structure and water solubility as influenced by the side chain structure could be correlated with auxin activity. His conception was that the action of auxin was something of a physical binding of some fatty material to some more aqueous phase.

Muir *et al.* (1949) gave a third suggestion that hormones of the phenoxy acid type may combine with some material (presumably proteinaceous) at the ortho position of the ring. They conceived that hormones react with some material in the cell at two positions. viz (i) at some position in the ring (ortho position in the phenoxy acids), and (ii) at the acid group of the side chain. On the basis of these Foster *et al* (1952) advanced a theory of hormone activity by two-point attachment and brought forward variety of kinetic evidences to support their hypothesis.

Theories of Enzyme Effects

It has been observed that growing tissues when treated with auxin show an increased activity, either directly or indirectly, of a number of enzymes. This fact has given rise to a theory of auxin action through enzyme mechanism. Northen (1942) found that auxins cause decrease in cytoplasmic viscosity and suggested that they bring about dissociation of the proteins constituents of the cytoplasm. This dissociation effect would increase water permeability, increase the osmotic value of the cytoplasm, and possibly also bring about increased enzymatic dissociation can sometimes activate enzymes and such dissociation effect would increase water permeability, increase the osmotic value of the cytoplasm, and possibly also bring about increased enzymatic activity. It has been shown by Hand (1939) that gentle protein dissociation can some times activate enzymes and such dissociation might increase the availability of substrates for the enzymes as well. As a consequence of this an increased respiratory activity and growth might follow.

Thimann (1951) postulated that auxins may act, not as enzyme activating agents, but as agents protecting growth enzymes from inactivation. He points out that organic acid metabolism is primarily responsible for growth. One of the enzymes (succinic dehydrogenase) in this system, which has a sulfhydril structure, is a critical link and may be very intimately involved in growth. Poisons which attack sulfhydril enzymes also antagonize auxin action and growth.

Bonner (1949) and Bonner and Bandurski (1952) have suggested that auxins may serve in some way to couple or mesh together the respiratory process with growth processes, and thus they act in a manner which would make the energy formed in respiration available to growth processes.

Theories of Osmotic Effects

Water uptake, serves as the basis of growth through increase in cell volume. The uptake of water could be due to changes in the cytoplasm itself, specially changes in osmotic value; or it could be due to changes in the properties of the wall and the cell membranes, particularly with respect to extensibility and permeability. Each of these two factors in water uptake has been defended as a possible mechanism through which auxins may bring about growth.

Czaza (1935) suggested that auxins may increase the osmotic value of the cell sap which would result directly in water uptake and growth. A basic limitation to Czaza's concept lies in the fact that growth is not necessarily associated with an increase in osmotic value. Actually van Overbeek (1944) and Hackett (1951) have been able to obtain growth with an associated decrease in osmotic value.

Commoner et al. (1942,1943) have proposed that auxin may influence the respiratory uptake of salt and the resultant increase in the osmotic value of the cell sap would cause water uptake and hence growth. Reinders (1942), however, has shown that auxin induced growth can occur in pure water and in the absence of salt uptake.

Reinders (1942) and Mackett and Thimann (1952) have collected evidence that water uptake in response to auxin treatment is dependent upon oxidative metabolism. Bonner et al, (1953) suggest that phosphorylation reactions provide energy for water uptake phenomenon.

Theories of Cell Wall Effects

Direct measurements of the effect of auxins upon the stretching properties of the cell walls have been made by Heyn (1932, 1933). He found that auxin application causes an increase in the flexibility and extensibility of cell walls. According to this, increased extensibility would result in a drop of wall pressure around the cell and would allow water uptake due to this simple drop in turgor pressure. This concept attributes growth primarily to the dynamic function of the cell wall.

Burstrom (1953) after a critical study of water uptake concluded that cell wall attains a plastic quality during growth and that water uptake only follows the changes in the cell wall. Thus water uptake is a consequence and not the cause of growth.

Toxic Metabolism

It is known that auxin, besides being growth promoters, can also cause inhibition of elongation. In shoots this inhibition requires high concentrations of auxin and is not prominent. In roots the regular response to exogenous auxin is an inhibition of elongation even at relatively low auxin concentrations. Very little is known concerning the mechanisms of this specific inhibition. According to one view it is conceived that the inhibition is due to excess auxin molecules inactivating the sites of auxin action and thus preventing maximum growth response. This mechanism although plausible, is supported by limited decisive evidences.

It may be concluded that the mechanism of auxin action remains yet unsolved, but a variety of promising lines of evidence seem to be emerging. The different angles from which the problem is being tackled may provide valuable lines of approach.

THE SYNTHETIC GROWTH HORMONE

A natural consequence of the discovery of auxin activity was the isolation and characterization of the auxin molecule. As soon as this was accomplished, an intensive search began for compounds chemically similar to IAA and with similar activity. Before long, the results of this search brought forth other indole derivatives, such as indole-3-propionic acid, indole-3-butyric acid and indolepyruvic acid all of which demonstrate physiological activity similar to that of IAA. Other compounds, similar in activity but not in chemical structure of IAA were also discovered. Of these, the more important ones are α- and β-naphthylacetic acids, phenylacetic acid naphthoxyacetic acid and phenoxyacetic acid. The structures of the above-mentioned compounds are given.

Molecular Structure and Auxin Activity

The chemical characterization of physiologically active compounds never fails to generate interest in the relationship between the structure of the compound and its physiological activity. It was interest such as this that led to the listing of certain minimal requirements needed by a compound for auxin activity. These requirements are:

1. An unsaturated ring system.
2. An acid side chain.
3. Separation of the carboxyl group (–COOH) from the ring (several exceptions).
4. A particular spatial arrangement between the ring system and the acid side chain.

The above requirements are minimal for auxin activity. However, the degree of substitution in the ring and side chain, nature of the ring (indole, phenyl, anthracene, etc.), and length of the side chain all are factors that will influence auxin activity.

Nature of the Ring System

After IAA had been isolated and characterized, it was soon found that the nitrogen of the indole ring was not essential for auxin activity. When either a carbon or an oxygen atom was substituted for nitrogen activity, although considerably diminished, was still observed. One might have anticipated this due to the fact that ring sizes ranging from the small phenyl ring to the relatively lare anthracene ring have been found in compounds having auxin activity. Nitrogen is not found in the phenyl or anthracene ring. There seems to be considerable evidence supporting the requirements that the ring be unsaturated. Activity decrease with hydrogenation of the double bonds of the ring, with complete cessation of activity occurring in the saturated ring.

It was not until the auxin activity of the phenoxyacetic acid series was discovered that the profound effect of the substitution of various groups onto the ring or side chain was truly appreciated. The nature of the group substituted and the location of the substitution was found to influence the activity of the compound. A striking example of this may be seen in the substitution of the chlorine atom at various positions on the phenyl ring of phenoxyacetic acid.

This was clearly demonstrated by Muir *et al.* (1949) when they experimented with the substitution of halogens and methyl groups in the 2, 4, and 6 positions of the phenyl ring. They found that substitution of both the 2 and the 6 positions with chlorine atoms resulted in complete loss of activity. However, substitution of the 3 or 4 position alone increased activity. Chlorine atoms attached to the 2 and 4 positions of the phenyl ring of the phenoxyacetic acid molecule gave the greatest auxin activity. Instead, 2, 4 - dichlorophenoxyacetic acid (2,4-D) is one of the most widely used synthetic auxins at the present time. The fact that there is a complete loss of activity when both the 2 and the 6 positions are chlorinated, led these investigators to hypothesize that the position on the phenyl ring adjacent to the point of attachment of the side chain is involved in the growth reaction.

Nature of the Acid Side Chain

According to the work of Koepfli et al. (1938) and others, the position and length of the acid side chain has a striking influence on auxin acitivity. Side chains that have the carboxyl group separated from the ring by a carbon or a carbon and oxygen given optimal activity. For example, the acid side chains of IAA and 2,4-D, two highly active auxins, fulfill these requirements.

As the length of the side chain in the phenoxyacetic acid series in increased, there is a falling off of activity. This drop is erratic, activity falling much lower when the side chain contains an odd number of carbons. That is, 4-dichlorophenoxybutyric acid (4 carbons) is more active than 2, 4-dichlorophenoxy propionic acid

(3 carbons). A possible explanation for the above phenomenon may be found in the work of Synerholm and Zimmerman (1947) and Fawcett et al. (1952). Apparently, side chains containing an odd number of carbons are metabolized in living tissues to the inactive phenol, where as the even-numbered chams are broken down to the active phenoxyacetic acid.

Substitution of different groups on to the side chain also affects activity. Thus, substitution of a methyl group on the carbon of the side chain of phenylacetic acid does not take away auxin activity. However, substitution of two methyl groups on the carbon eliminates auxin activity entirely. Later, when we discuss the mechanism of auxin action, we will learn why the substitution of "bulky methyl" groups on the acid side chain inhibits auxin activity.

Hydorxyl substitution in the side chain also can eliminate auxin activity. For example, substitution of a hydroxyl group or an alcohol group on the carbon of phenyl acetic acid produces two inactive derivatives.

Earlier, it was thought that separation of the carboxyl group from the side chain was essential for auxin activity. However, many exceptions to this rule have been found. For example 2,3, 6-trichlorobenzoic acid exhibits a strong auxin activity.

Spatial Arrangement

The spatial relationship between the ring and side chain is an important factor in the activity of an auxin molecule. For example, cis-cinnamic acid demonstrates good auxin activity in the *Avena* straight growth test, and its *trans* isomer does not. Veldstra (1944) suggested, after a study of the relation between structure and activity, that in order for a molecule to have auxin activity the –COOH group and the ring should lie in different planes. This theory was supported by observation of the activity of *cis* and *trans* forms of tetrahydronaphtylideneacetic acid and the above *cis* and *trans* forms of cinamic acid.

Antiauxins

If one assumes that the chemical configuration of a compound is responsible for its effect on physiological processes, then we must also recognize that there can be an interference with such action by similar, but not identical, compounds . Many antiauxins have been discovered and, generally, these combination with auxin, inhibit its acitivity.

Actually, the term antiauxin should define only those compounds that will complete with auxin for a reactive site in the growing cell. What do we mean by a reactive site? Because of the relationship of molecular structure and configuration with degree of physiological activity, most theories on auxin action are based on the attachment of the auxin molecule to some substance in the cell (e.g. protein). This complex (bound auxin)can then induce auxin activity. True antiauxins, then, are compounds which, because of the molecular similarity to auxin, would become attached to reactive site, thus neutralizing them for action in growth . However, if true competition for reactive sites is involved here, increasing applications of auxin should eventually swamp the reactive sites with auxin molecules and overcome the antiauxin effect.

There is general agreement that in order for an auxin molecule to be active, it must make a *two -point* attachment at a reactive site. In addition, it is thought that the two points of attachment are made through a position on the unsaturated ring and by the carboxyl group of the side chain. In the phenoxyacetic acids, the position on the ring through which attachment is made is the *ortho* position.

Based on the two-point attachment theory, McRae and Bonner (1953) have drawn up a rigorous classification for true, antiauxins. Using 2, 4-D, a synthetic auxin, as representative molecule, they have shown where "analogs" of the 2, 4-D moleucle containing some, but not all, of the structural properties of the auxin are capable of competing for reactive sites. The antiauxin makes a cone-point instead of the two-point attachment necessary for growth action. McRae and Bonner have listed three ways by which a modification of the 2,4-D molecule can lead to antiauxin activity.

1. Elimination of the essential carboxyl group.
2. Elimination of the essential reactive *ortho* group.
3. Elimination of proper spatial relationships between the ring and the carboxyl group as by introduction of bulky groups in the side chain.

Although not as popular as the two-point attachment theory, a three-point attachment theory has been proposed as necessary for auxin activity. According to this theory, an auxin molecule in order to be active must contain the following necessary structural, properties; an unsaturated ring, a carboxyl group, and at least one α-hydrogen. A further stipulation is that all three must be correctly oriented in space with each other. Of the isomers of 2, 4-dichlorophenoxy-α-propionic acid, only the "+" form is active. The "–" form is not properly oriented, thus failing to accomplish the three -point attachment necessary for auxin activity.

Contact is made by the auxin with the reactive site simultaneously at three positions on the auxin molecules. If only one site or even two positions are occupied, no activity ensues. In fact, molecules that occupy only one or two of the positions on the reactive site can be considered antiauxins.

A type of antiauxin which has not, as yet, been considered is the compound having weak auxin activity. A weak auxin can make the necessary two-point (or three-point) attachment and initiate a stimulation of growth. However, this stimulation is small, and at the same time active sites involved with the weak auxin cannot be occupied by a strong auxin. An example of a weak auxin having antiauxin characteristics is phenylbutyric acid.

Kinetics of Auxin Activity

A valuable contribution to the study of auxin-induced growth reactions has been the application of the Lineweaver-Burk kinetic analysis of competitive inhibition. McRae and Bonner (1953) demonstrated that auxin-induced growth reactions could be treated by the methods of classical enzyme kinetics.

It is generally accepted that in enzyme reactions, an intermediate complex is formed between the substrate and enzyme and that this complex is formed at an active site on the enzyme. The complex is converted to the enzyme and reaction products.

$$E + S = ES \rightarrow E + P$$

Using the above scheme, McRae and Bonner demonstrated that stimulation of straight growth of *Avena* coleoptile sections by auxin could be mathematically analyzed. In their scheme, enzyme (E) is the auxin receptor, substrate (S) is the auxin applied, complex (ES) is the receptor-auxin attachment, and the product in this case is growth. Let us rewrite the above equation using A for auxin, R for receptor, RA for the intermediate complex, and G for growth.

$$R + A = RA \rightarrow R + G$$

Now, in classical enzyme study, a competitive inhibitor is thought of as a compound that will compete with the normal substrate of an enzyme for active sites on that enzyme. The formation of an enzyme-inhibitor complex may be illustrated in the following manner.

$$E + I = EI$$

The fact that the formation of EI is reversible is important since this allows for competition by the substrate with the inhibitor for active sites. Therefore, by increasing the substrate, inhibition by a competitive inhibitor may be overcome.

The effect of a competitive inhibitor may be observed in a decrease in the rate of an enzyme reaction. However, if the substrate concentration is increased until all active sites on the enzyme are "swamped," then maximum velocity (V_{max}) or rate will be obtained. This maximum velocity will be identical to that of the same reaction without a competitive inhibitor. In other words, increasing the concentration of a substrate decreases the amount of inhibition, and conversely, decreasing the substrate concentration increases inhibition.

We have mentioned that antiauxins compete with auxins for active sites on an auxin receptor or growth center. This situation,of course, is analogous to the concept of competitive inhibition in enzyme study.

Now, using the Lineweaver-Burk plotting method, one can measure the velocity of an auxin reaction (in this case, rate of growth) and, at the same time, calculate the effect of an antiauxin on this velocity. By plotting the reciprocal of the velocity (I/V) of the auxin reaction against the reciprocal of the concentration of auxin (I/[A]) applied, one obtains a straight line relationship. Maximum velocity (V_{max}) may be obtained by extending this line to the ordinate, the intercept being I/V_{max}.

Note that the concentration of inhibitor will affect the slope of the line, but not the intercept. A mathematical analysis of the interaction of 2,4-D and the antiauxins 4-chlorophenoxyisobutyric acid, 2,6-dichlorophenoxyacetic acid, and 2, 4- dichloroanisole has been under taken. The compound 4-chlorophenoxyisobutyric acid is an antiauxin because of bulky methyl groups in the side chain interfering with attachment of the carboxyl group to the auxin receptor. 2,6-Dichlorophenoxyacetic acid owes its antiauxin properties to block age of the reactive *ortho* positions on the ring by chlorineatomes. 2, 4-Dichloroanisole has to carboxyl group and therefore cannot make the necessary two-point attachment.

Inactivation of Auxin

Just as the production and subsequent physiological effects of auxin have a profound influence on plant development, the inactivation of auxin appears also to be of significance in this respect. For example, the inactivation of auxin is important in phototropisms, control of cell elongation, and in the aging of plant tissues. We will discuss the mechanisms involved in the destruction of auxin and the influence of its destruction on cell elongation and aging.

Mechanisms of Auxin Inactivation

A preponderance of literature has developed on the mechanisms of auxin inactivation since the isolation in 1947 by Tang and Bonner (1947) of an enzyme capable of oxidizing IAA. This enzyme is now called *IAA oxidase.* Recently an IAA oxidase was extracted from tobacco roots and partially purified. Although IAA oxidase represents one way for a plant to destroy auxin, other natural means of auxin inactivation have been discovered. However, two systems for the destruction of auxin in the plant appear to dominate, and these are (a) enzymatic oxidation and (b) photo-oxidation.

Enzymatic Oxidation

Enzyme systems that oxidize IAA have been found in several plant tissues. However, as is generally true, one system is usually studied in much more detail than the others, and in this case it is the enzyme system present in extracts of etiolated pea epiçotyls. It appears that in this system a flavoprotein must be present, which gives rise to hydrogen peroxide. The oxidation of IAA by hydrogen peroxide is catalyzed by a peroxidase to yield some inactive product, probably indolealdehyde. In the inactivation of IAA by this system, 1 mole of O_2 is consumed for every mole of IAA inactivated and CO_2 released.

In addition to indolealdehyde, other breakdown products have been suggested. However, it is generally accepted that the logical end product of IAA oxidation is indolealdehyde. As mentioned before, IAA oxidase systems have been found in many plants, and in several cases these have differed from the original IAA oxidase system found in the pea plant. Perhaps in these, different end products of IAA oxidation are obtained.

An inverse relationship between IAA oxidase activity and IAA content in the plant has been found. That is, where IAA content is high, IAA oxidase activity is low and vice versa. Meristematic regions that have a high auxin content have been found to below in IAA oxidase activity. The root, which is generally thought to below in auxin content, has been found to be generally thought to be low in auxin content, has been found to be high in IAA oxidase activity. In fact, Galston (1956) has found (at least in the pea plant) that as cells age, their IAA oxidase activity increases, and their auxin content drops.

Increased capacity to inactivate IAA has been demonstrated by young plant tissues treated with synthetic IAA or analogs of the IAA molecule. It appears, then, that IAA is capable of inducing the formation of the enzyme that destroys it. This is particularly interesting since IAA, which initiates growth, also puts into action the mechanisms leading to the termination of growth. Galston (1956), a leading figure in research on plant growth, has this to say: It seems possible that the decreased sensitivity to auxin of older cells is a consequence of their higher IAA oxidase activity, which in turn is a consequence of prior induction by IAA. According to this scheme,. the administration of IAA to a young cell not only initiates growth, but also sets in motion a chain of events leading to the diminution and eventual culmination of growth.

Photooxidation

It has long been known that IAA can be inactivated by ionizing radiation. Skoog (1934; 1935) demonstrated that rapid inactivation of pure IAA takes place when it is subjected to x-and gamma-radiation. He also noted that little, if any, inactivation takes place in a nitrogen atmosphere, suggesting that inactivation is due to oxidation by peroxides formed during irradiation. There is some evidence that only a small amount of IAA is inactivated or oxidized in this manner, most of the detrimental effect of this type of irradiation to IAA being of an indirect nature. For example, Gordon (1956) has claimed that the major effect of ionizing radiation on auxin metabolism may be found in the destructive effect of the radiation on the enzyme system converting tryptophan to IAA.

Ultraviolet light also inactivates IAA. This might have been predicted because of the ring structure of the IAA molecule, which absorbs to some extent in ultraviolet (maximum absorption at about 280 mμ). Here, there is a direct effect on the IAA molecule due to the absorption of ultraviolet light. Determinations of auxin content before and after irradiation with ultraviolet light have shown that this type of irradiation reduces auxin levels in plants.

GIBBERELLINS

The gibberellins are tetracarbocyclic compounds whose small amounts profoundly stimulate the growth of many plants.

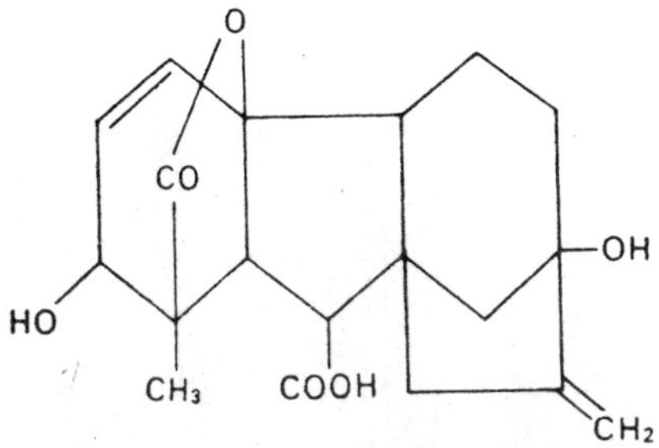

Fig. 7.9. Gibberellic acid.

Gibberellins were originally regarded as unusual and interesting fungal metabolites, but their importance in the regulation of plant growth has recently been emphasized by the widespread occurrence of gibberellin-like substances in the plant kingdom. The chemical identification of some of these has been established. The discovery of gibberellins is very interesting and rather accidental. These growth regulating substances were discovered as a direct consequence of interest in the "bakanae" disease of *Oryza sativa*, a disease which had serious economic effects on the rice industry in Japan. The disease has two important symptoms:

1. seedlings assumed long, thin, and pale green appearance than those of healthy plants, and
2. mature plants, sometimes taller than their healthy counterparts often failed to set fruit.

The "bakanae"symptoms were first described by the plant pathologist S.Hori in 1898, although its technical account appeared in the literature much later. Early in the twentieth century, extensive studies by Japanese plant pathologists conclusively showed the "bakanae"disease to be resulting due to certain strains of a species of fungus now known as *Fusarium monilliforme* Sheld (*Gibberella fujikuroi*). Sawada (1912) suggested that the seedling symptoms of the disease could be due to a "substance" secreted by the fungus. This was experimentally proved by Kurosawa in 1926. He demonstrated the appearance of bakanae symptoms in *Oryza sativa* fungus. This observation attracted little attention outside Japan, but the work on isolation and purification of the growth-promoter was initiated. Progress was slow but eventually, in 1939, Yabuta and Hayashi isolated a crystalline 'A'.

More recently Japanese chemists have suggested that this substance is mixture of several different growth promoters collectively known

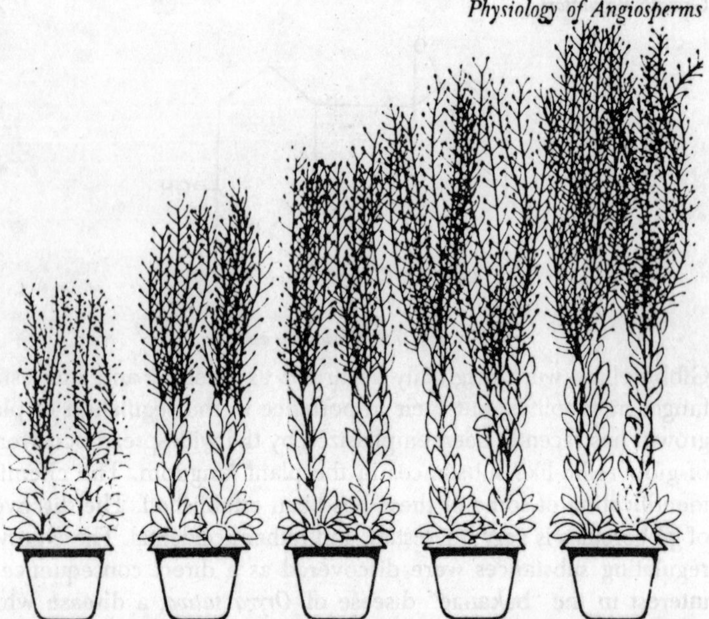

Fig. 7.10. Primrose require long days to flower. Plants were treated with increasing amount of Gibberellic acid, all plant flowered without long day conditions.

as gibberellins. Pure gibberellin was first obtained in England. Many years after the description of gibberellin 'A' in Japan, interest developed in other countries. All the Japanese work, from Kurosawa's observations of 1926 to the isolation of an active material by Yabuta and Hayashi in 1939 was promptly abstracted in "Chemical Abstracts" and the "Review of Applied Mycology".

In attempting to repeat Japanese work, Curtis and Cross isolated, in 1954, a substance obviously chemically and physiologically similar to the Japanese gibberellin 'A', but nevertheless distinct, they called this substance gibberellic acid. Shortly afterwards Stodola, in United States, described a new gibberllin which he called gibberellin 'X', this was found to be identical with gibberellic acid. Several pure gibberellins are now known. They are probably fairly closely related chemically and all have similar effects on plant growth.

Terminology

Following the suggestions of Phinney and West (1960), two terms, gibberellin and gibberellin-like are maintained. The former term is restricted to substances defined both by biological and chemical properties whereas the latter term defined by biological properties only. The term, gibberellins will therefore be reserved for substances biologically active in stimulating an increase in the size of a plant organ or parts thereof and known to consist chemically of a carbon skeleton identical with,or closely related to, that of gibberellin 'A_3' (gibberellic acid). The five gibberellins identified are : gibberellin 'A_1' (GA_1), gibberellin 'A_2' (GA_2) gibberellin 'A_3' (GA_3), gibberellin 'A_4' (GA_4), and gibberellin 'A_5' (GA_5).

Assays

Depending on the physical or chemical properties of the gibberellins rather than on their physiological activities, a number of qualitative and quantitative assays have been developed.

Physico-chemical assays

With the use of filter paper chromatography a variety of detection methods have been employed to locate gibberellins.

Fig. 7.11. A-B–Henbane plants which usually require a period of cold treatment and long days for flowering-flowered when they were treated with Gibberellic acid as in B (No low temp. or long day were required), C-D–Carrot plants require low temperature for flowering, C–Plant without low temperature but with gibberellin spray flowered, D–Plant provided with low temperature.

Gibberellin A_3 is defected by a characteristic fluorescence developed in sulfuric acid. The substance, by this method, could be detected in amounts as low as one microgram. The fluorescence of the fluorogen formed from gibberellin 'A_3' in sulfuric acid solution has been used as the basis for a quantitative analysis. A linear calibration curve is obtained for gibberellin 'A_3' concentrations in the range of 0.00625 to 3.2 micrograms per ml. The amount of fluorescene formed from a sample is subject to a number of variables which must be controlled to get reproducible gibberellins, GA_1, GA_2, GA_4 and GA_5. Other spray reagents used include 0.5 percent potassium permanganate reagent and an acidic periodate-permanganate reagent.

A number of useful isotopic labelling methods, for the assay of small amounts of gibberellins present in relatively crude mixtures have been developed by Baumgartner (1958) and Banumgratner *et al.* (1959). Selected absorption bands in the infrared region has been used for the quantitative determination of gibberellins and a polarographic method.

Biological assays

Angiospermous plants in which the dimensional change of the shoot, or parts thereof occurs, serve as a measure of response in these assays. Floral induction and hastening of flowering have also been used to a limited extent for the detection of gibberellin-like substances. Generally, the assays which employ intact plants are more specific than those which employ the growth responses of plant sections in culture. The various bioassays used for the determination of gibberellins and gibberellin-like substances can be summarized.

Gibberellin-Like Substances from Flowering Plants

The presence of the giberellins in *Fusarium* has already been discussed. The first definitive evidence for the presence of gibberellin-like substances in organisms other than *Fusarium moniliforme*, came as a result of the use of gibberellin-type of growth response in bioassays, assays that could be shown to be specific for gibberellins only. Several worker groups have reported the widespread occurrence of gibberellins in flowering plants.

Gibberellin-like substances have been reported from 64 genera reporting 15 different families of angiosperms, from one member of the mosses, from two genera of fungi and also from two members of Actinomyces. These gibberellin-like materials have been reported

Table 7.2. Species of plants used as assays for gibberellins and gibberellins-like substances.

Plant species	Plant part	Specificity
1. *Avena sativa* L.	leaf section	non-specific
2. *Perilla ocymoides* L.	intact seedling	specific
3. *Pharbitis Nil chois*	intact seedling	specific
4. *Phaseolus vulgaris* L.	intact seedling	non-specific
5. *Pisum sativum* L.	intact seedling	specific
6. *Oryza sativa* L.	intact seedling	specific
	leaf section	non-specific
7. *Rudbeckla bicolo* Nutt	rosetted plants	specific
8. *Triticum vulgare* Vill	leaf sections	non-specific
	excised coleoptile	non-specific
9. *Zea mays* L.	intact seedling	specific

from the root and all organs of the shoot, and, from both young and old tissues of flowering plants. Macmillan and Suter (1958) have isolated gibberellin A_1 in small quantities from immature seed of *Phaseolus vulgaris.* West and Phinney (1959) extracted a substance, bean factor I, from immature seed of *Phaseolous vulgaris* and which subsequently was shown to be identical with gibberellin A_1, Kawarada and Sumiki (1959) have isolated gibberellin A_1 from "water sprouts" of *Citrus unshuii.* West and Phinney (1959) obtained a second substance, bean factor II, form extracts of immature seed of *Phaseolus vulgaris,* which had gibberellin like biological properties but differed in chemical and physical properties from the reported fungal gibberellin. Macmillan, Seaton and Suter (1959) also obtained a second gibberellin, which they name gibberellin A_5, from their extracts of *Phaseolus multiflorus* seed. A comparison of the infrared spectra of gibberellin A_5 and bean factor II, and their methyulesters, has shown that these substances are identical. The name gibberellin A_5 has been accepted by both groups. gibberellin A_3, the gibberellin produced in largest amounts in most fermentations has not been isolated from flowering plants.

Chemistry of Gibberellins

Structural Studies on the Fungal Gibberellins

The two fungal gibberellins, gibberellin A_3 and gibberellin A_1 are structurally known as reported by the British group at Akers Research :Laboratories of ICI. The results of more recent investigations of the Tokyo group on gibberellin A_1 and gibberellin A_2 fractionated from the gibberellin A mixture, have led to proposals IB and IIB for gibberellin A_3 and gibberellin A_1 respectively. These differ from the above proposal only in the positions of attachment of the lactone ring in A.

The relationship between gibberellin A_1 and gibberellin A_3 has been established. Grove *et al.* (1958) could partially hydrogenate methyl gibberellate ($C_{20}H_{24}O_6$) and separate from the mixture of products two isomeric methyl ester ($C_{20}H_{26}O_6$). One of these methyl-α-dihydrogibberellate, was identical with the methyl ester of gibberellin A_1. Since gibberellin A_1 still possesses an exocyclic methylene group as indicated by ozonization studies the two hydrogens must have added to saturate the A ring. A few of the products which have been derived by chemical treatment of gibberellin A_2 and gibberellin A_1 can be summarized.

Stodola et.al (1957) and Cross (1954) indicated the molecular formula $C_{19}H_{22}O_6$ for gibberellin A_3. The functional groups in this were identified as a carboxyl, two ethylenic double bonds, two alcoholic hydroxyl groups (one secondary and one presumably tertiary), a saturated lactone, and probably one methyl group. This indicated a compound with four carbocyclic rings. Acid hydrolysis yielded a number of products, such as allogibberic acid (identical with gibberellin B of Japanese) and gibberic acid.

Cross et al. (1956 ,1958) assigned the structure 'V' to gibberic acid. The Japanese group obtained the hydrocarbon, gibberene, as a product of the selenium catalyzed hydrogenation of gibberellin A or gibberellin B or gibberic acid. Yabuta et al. (1941) identified this hydrocarbon as a substituted fluorene. Mulholand and Ward (1954) also isolated gibberene as a hydrogenation product of giberric acid and showed it to be 1,7 dimethyl-fluorene. These probable presence of a substituted, perhydrofuorene nucleus is gibberic acid. Gibberic acid was also converted by Cross et al (1958) to methyl -1, 7-dimethyl fluorene-9-carboxylate by appropriate series of reactions. This has established the position of the carboxyl group. By a step-wise degradation procedure gibberic acid was converted to the methyl ester of a tetracarboxylic acid by Cross et. al.(1958).

Cross et al. (1956), and Mulholand (1958) assigned the structure IV to allogibberic acid. Like gibberellin A_2, it gave formaldehyde of ozonolysis indicating the presence of a terminal methylene group. Allogibberic acid was converted to 7-hydroxyl-1-methylfluorene by a series of reactions including a selenium dehydrogenation. This located the hydroxyl group substitution on the nucleus.

Gerzon, Bird and Woolf (1957) assigned structure IIIa or IIIb to gibberellenic acid–another product obtain from gibberellin A_3. Gibberellenic acid is readily converted in acidic solution to allogibberic acid, which in turn is readily converted to gibberic acid, an end product of the acid hydrolysis.

The establishment of structure of allogibberic acid and gibberic acid also fixed the structure in the B, C and D rings of gibbereilin A_2. It still remained to be established how the lactone, secondary hydroxyl and ethylenic double bonds were accommodated in the ring A. The position of the secondary hydroxyl was indicated by the conversion of the methyl ester of gibberellin A_1 through a series of intermediates to 2- hydroxyl 1, 7-dimethyl fluorene or 2-hydroxyl-1- methylfluorene. There seems at present no satisfactory way, however, of reconciling all the reported degradation products with one structure of gibberellin A_2 and gibberellin A_1. Structure has been proposed by Kitamura et al. (1958, 1959) for gibberellin A_2 and structure for gibberellin A_4. There structures are similar to that proposed for gibberellin A_1 (ring A is shown in the form proposed by the Japanese) except for the position of substitution of the tertiary hydroxyl and the absence of unsaturation in A_2 and the absence of the tertiary hydroxyl in A_4.

Structural Studies on Gibberellins from Angiosperms

The chemical nature of the active principles in angiosperms has been determined in only a few instances. Some of the most important sources from which the gibberellins have been isolated in angiosperms has already been mentioned. Structure has been assigned to gibberellin A_5 by MacMillan *et al.* (1959). Thus, gibberellin A_5 is dehydrogibberellin A_1 in which the secondary hydroxyl grop is eliminated from the ring A. A key compound in the structural proof was the keto-acid formed from gibberellin A_5 on acid treatment. Thus at least two gibberellins, one identical to a fungus-produced gibberellin and the other very closely related to the fungal gibberellins chemically, have been found to occur naturally in flowering plants.

The Relation of Chemical Structure to Biological Activity of Gibberellins

In an assessment of the relation of chemical structure to biological activity of gibberellins many factors influence the interpretation of results. These include (i) the types of assay methods used; (ii) the method of application of the test material, and (iii) accompanying permeability problems and the range of concentration of the test material. Studies with the naturally occurring gibberellins have shown gibberellin A_3 to be the most active followed by gibberellin A_1, gibberellin A_4 and gibberellin A_2 in decreasing order of activity in most assay used. But this is not true always because the activity of gibberellin A_4 stimulating growth of some cucurbits at levels at which the other gibberellins are inactive, has been found. Also depending on the particular dwarf mutant used for assay, gibberellin A_5 is as effective as gibberellin A_2 or less than 10 percent as active as gibberellin A_1.

Gerzon et al. (1957) showed that gibberellenic acid formed from gibberellin A_3 in acid solution, is devoid of biological activity. Allogibberic acid and gibberic acid which are formed from gibberellin A_3 on more vigorous acid treatment are likewise devoid of biological activity. Besides, gibberellin C, formed on acid catalyzed rearrangement of gibberellin A_1, retains some activity. The isomeric products formed from gibberellin A_1 and gibberellin A_2 by mild alkaline treatment are biologically inactive. Under some condition which do not open the fact one ring, gibberellin A_1 is inactivated; this is believed to be due to the formation of an primer (pseudogibberellin A_1) in which the secondary hydroxyl group in the ring A is inverted from an axial to a more stable equatorial configuration. On mild alkaline hydrolysis of gibberellin A_3, the diacid results by an allylic rearrangement of the hydroxyl group formed on opening the lactone ring.

The free acid and salts both, of gibberellin A_3 are biologically active. It has been found that conversion of the carboxyl group of gibberellin A_3 to the methyl, ethyl, butyl, or acetyl esters results in a complete loss of biological activity. Generally, those products derived from the gibberellins which involve modification of the A or B ring structures are inactive. Sumike and J Kawarada (1941) found the derivative of gibberellin A_2 in which the alcohol is oxidized to a Keto Group, to be inactive. Many of the products derived from the gibberellins which involve structural modification only in the C or D rings, such as gibberellin C, are active.

Biosynthesis of Gibberellins

The broad outlines of the gibberellin biosynthesis have been indicated but very little detailed knowledge has been gained. Cross et al. (1958) advanced "biogenetic grounds" as one of their reasons for favouring the carbon skeleton they proposed for gibberellin A_3, since it could be considered as an "isoprenoid" (isopentane) derivative. Birch et al. (1958), and Birch and Smith (1959) have proposed a general route from acetate and mevalonate (β-δ-dihydroxy-β- methyl-valerate) to gibberellin A_2 *via* diterpenoid intermediates. Some key features of this proposal may be summarized. Mevalonate (B) which can be biosynthesized from acetate, has been shown to be a precursor of sterols and terpenes.

Condensation of four molecules of a mevalonate derivative in a "head- to-tail" fashion leads to an acyclic diterpene with the carbon skeleton on C. The cyclization of the acyclic precursor leads to a tricyclic diterpene of carbon skeleton D.Subsequent modification of the skeleton to that of gibberellin A_3 (E) would occur by (1) elimination of the C_{17} methyl group, (2) a ring contraction of the B ring, and (3) a rearrangement involving the C ring and its attached vinyl group to produce the phyllocladene type C-D rings of gibberellin A_3.

This general proposal has been supported by the labelling studies. The previous figure shows one of the ways in which gibberellin A_3 would become labelled from acetate-1-C^{14} or mevalonate -2-C^{14}. Zweig and Cosens (1959) have also presented C^{14} labelling evidence for the incorporation of both the methyl and carboxyl carbons of acetate into gibberellin A_3 by the fungus. Little, direct information bearing on the biosysnthesis of gibberellins in flowering plants is available at the present time. It is presumed that the route in the fungus and in flowering plants will be very similar.

Anatomical Effects of Gibberellins

It is now well established that changes in size and form of plants take place due to gibberellin treatment. Hence it raises the problem of whether these effects are the result of changes in the net cell number, net cell size (including form of the cell) or combinations of the two. In general, it was showed by Japanese that increased cell length rice seedlings. Nevertheless since some of the early studies showed increased cell elongation to be insufficient to account for all of the elongation observed at the organ level, increased cell number was also implicated but only by inference and only to a minor degree. The discovery by yabuta and Hayashi (1939) that gibberellin increased the "reproductive rate"of *Lemna paucicstata* gave a signal that the effect must have been associated with increases in cell number.

It is now realized that gibberellins may markedly increase both number and length of cells, the effect, however, depending on the nature of the plant material and the conditions under which the material is grown. Lang (1956) pointed out that stem elongation in gibberellin treated *Hyoscyamus niger* must be due to an appreciable increase in cell number since the internodes are essentially absent in the non-treated iosetted forms. In 1956 Lona reported that cell length increases would account only for about 13 percent of the elongation of *Perilla ocymoides* stems following gibberellin application. He thus concluded that gibberellin treatement must result in an appreciable increase in cell number in the stem of this plant. Sachs and Lang (1957) and Sachs, Breitz and Lang (1959) have shown a very marked increase in the frequency of dividing cells in gibberellin induced elongation, of *Hyoscyamus niger* and *Samolus parviflorus.* Bradley and Crane (1957) showed an increase in cell number in the meristems of spur shoots of gibberellin treated *Prunus armeniaca.* Gundersen (1958) has observed increase in both cell number and cell lengths in certain of the internodes of *Begonia*, and both cell number and cell lengths are factors in the gibberellin induced shoot growth of *Phaseolus vulgaris.* This is also true for petiole elongation of *Fragaria* sp and seedling elongation of *Zea mays* and *Pharbitis nil.*

Increased cell number is indeed a reflection of an increase in mitotic activity. In angiosperms this increase due to gibberellin treatment is most marked in meristems. Differential growth is also a predominant feature of the gibberellin response at cellular level. The planes of mistoses are such that cell progenies are almost exclusively formed in one direction, followed by cell elongation in the same direction. The response in cell number and cell length would suggest that the effect of gibberellins may be to control a physiological mechanism common to both. While it appears that mitotic activity can precede cell elongation there are cases of gibberellin-induced elongation of pollen tubes which occur without any preceding flush of mitotic activity. Haber and Luippold (1960) have shown that the gibberellin-induced growth of *Triticum vulgare* seedlings, which come from irradiated grains, must be due to increases in cell lengths only, since the irradiation completely inhibits mitosis. Thus it appears that the two types of responses at the cellular level can occur independently of one another. It is possible that the response of cell to the application of gibberellin will depend on the physiological age of the cell, i.e., enhanced cell division (mitotic activity) in meristems, primarily cell elongation in regions of elongation, and no response or very limited response in regions which consist of differentiated cells.

Gibberellin-Induced Growth

A number of types of growth responses now known to be associated with the gibberellins have been reported. Some of the "bakanae" symptoms have already been described. Here an attempt will be made describe some of the general growth responses induced by gibberellins.

Among the various growth responses induced by gibberellins shoot growth is the most apparent response of flowering plants. At low dosage level the plant size is generally increased, while little or no changes in form take place. At higher dosage levels the response may become one of growth, primarily, in the dimension of normal elongation of the plant. excessive treatment with gibberellins may result in marked differential growth to produce plants with leaves and long and thin stems. The ability to respond is dependent on the genotype of the plant and the environment in which the plant is grown. Field grown plants are often more responsive to gibberellin treatment when grown under environmental stresses. Gibberellin treatment may result in inhibition of lateral buds and thus lateral branching. Abscission has also been reported to be affected by gibberellin treatment. The increased stem thickness often seen is indeed a reflection of the stimulation of the cambium and its immediate cell progeny.

The type of leaf response is also associated with the developmental pattern of the leaf. Thus leaves of grasses with intercalary meristems respond mainly by elongation. Leaves of dicotyledons may respond by increases in area of the blade. Both acceleration and retardation of abscission have been reported.

Generally roots of intact plants show inhibition or no response as a result of gibberellin treatment, although exceptions are found in seedlings where elongation of the radicle has been reported, for example, among the gymnosperms, *Pseudotsuga menziesii* and *Pinus lambertiana.* Rooting of cuttings has been observed to be inhibited and certain cases of modulation of roots are inhibited of a consequence of gibberellin treatment.

Physiological Role of Gibberellins

The specificity of response of many flowering plants to exogenous gibberellin, and the widespread occurrence in flowering plants of gibberellin-like substances of which some have been isolated and chemically identified as true gibberellins, establishes the fact that gibberellins may be regarded as native plant growth regulators. Critical evidence for a causal relationship between endogenous gibberellin and growth has come from the positive correlation of certain types of growth with the level of endogenous gibberellin-like substances. Although synergism between gibberellins and auxins have been demonstrated for a number of cases, nevertheless the mechanistic interpretations of these data are questionable beyond the idea that numerous growth regulators must be present in optimum amounts for maximum growth of an organism. At present, however, the more convincing examples of growth which may be dependent on a gibberellin mechanism are those which exhibit a specified response, i.e. they respond to gibberellins but not to auxin, other known growth regulators.

Shoot Growth

Many cases of appreciable shoot growth resulting specifically from gibberellin treatment are now known. These include; the bolting and flowering responses of numerous photoperiodic and cold-requiring plants, normal growth of certain single gene dwarf mutants, elongation of red-light-inhibited plants, and possibly the breaking of certain types of dormancy and shortening of the after-ripening time required by certain seeds. Generally the effects seem to be not of differentiation of tissues but more of enhanced growth of preexisting cell types, with changes in form being consequences of differential growth. Whenever examples of primordial initiation are recorded, they are apparently indirect consequences of a gibberellin-induced general growth stimulation.

Bolting and flowering

Most rosetted long-day plants and rosetted cold-requiring winter biennials bolt and flower in response to exogenous gibberellin,.and these effects can be brought about under an environment which would normally maintain the rosetted habit of growth. With appropriate gibberellin dosages the bolted and flowering plants appear very similar to those induced by photo induction or cold treatment. At limiting dosages of gibberellin, however, it is possible to separate shoot elongation (bolting) from floral differentiation so that bolting will occur without flowering. Such experiments lead us to the view that the flowering response is only an indirect effect of the gibberellin treatment, as a consequence of increased growth of the shoot; the endogenous factors than gibberellins are elaborated and ultimately result in the differentiation of floral primordia. That the action of gibberellins in flowering is an indirect one is also evidenced by the fact that short-day plants do not flower even though the gibberellin treatment may cause appreciable stem elongation. It has been reported that the flowering response can be enhanced by gibberellin treatment in *Xanthium pennsylvanicum* under marginal day lengths which by themselves will produce some flowering, there are, nevertheless, no examples of short-day plants grown under long-day conditions, shown floral induction following gibberellin application, and in some cases gibberellin treatment has been found to reduce flowering in short-day plants under short-day conditions. Thus it is apparent that gibberellin will replace the long-day requirement for flowering but not the short-day conditions. Thus it is apparent that gibberellin will replace the long-day requirement for flowering but not the short-day requirement, and the gibberellin-treated plants show appreciable stem elongation just as do those plants growing under long-day conditions.

Studies of Lang (1956,1957) that there exists a correlation between levels of native gibberellin-like substances and the presence or absence of bolting in biennial *Hyoscyamus niger*, have also been extended to other plants. By use of *Zea mays* bioassay, several gibberellin-like components are found in appreciably higher amounts than in non-bolting plants. Harada and Nitsch (1959) and Nitsch (1959) have also found a positive correlation between amounts of gibberellin-like substances and bolting in the cold-requiring plant *Chrysanthemum morifolium. Ram* cv. *Shuoka*, and the long-day plant *Rudbeckia speciosa* Wenderoth. They have also demonstrated increases in auxin-like substances as a result of photoinduction and cold treatment. However, since only exogenously applied gibberellins, and not auxins, induce bolting without photoperiodic or temperature induction, it may be inferred that the increase in endogenous auxin area consequence of the increased level of the native gibberellins. The accumulated evidence indicates that gibberellin may be the primary limiting factor in the bolting of at least some long-day and cold-requiring rosetted plants.

Dwarfism

It has been found that gibberellin treatment does not always lead to stem elongation, since many cases have been recorded where the process is actually reversed by gibberellin treatment. Reduced growth results from a shortening of the internodes rather than a decrease in number of internodes. As a result, the gibberellin treated dwarfs become essentially indistinguishable from the tall or non-treated forms, by the elongation of internodes rather than by the increase in the number of internodes. In considering possible mechanisms to explain genetic dwarfism, a clear distinction should be made between the use of the general term "genetic dwarf: and the specific term "single gene dwarf". In cases of single gene dwarfism, the gene responsible for dwarfism, may in someway block a single step in a metabolic pathway. The theoritical unitary control in a single gene mutant suggests that a unitary limiting factor may be responsible for the dwarf habit of the growth. In contrast genetic dwarfs can also include cases which result from several mutant genes (multiple factors). A multiple phenomenon may have a multiple control at the physiological or biochemical level.

It has been reported by Brian and Hemming (1955) that certain dwarf cultivars of *Pisum sativum*, *Vicia faba* and *Phaseolus multiflorus* respond to gibberellin A_3 to become practically indistinguishable from the tall cultivars. At the same time the tall cultivars showed little or no response to exogenously applied gibberellin. It is now known that the ability of a genetic dwarf to respond to gibberellin can be under the control of a single gene. Such a unitary control of the gibberellin response has been demonstrated for the gene in *Pisum sativum* the *d* gene in *Lolium perenne* and d_1, d_2, d_3, d_5 and an_1 genes in *Zea mays.* The dwarf mutants of *Zea mays* have been usefully used for physiological studies of gibberellin action. Out of the ten mutants studied, nine were simple recessives and one a simple dominant. Five have been found to respond to gibberellin treatment by normal growth, other five showing little or no growth response. None of these mutants respond to other growth regulators or to variations in the nutrient medium. The specificity of response of the responding mutants suggests that each gene is in some way controlling a different step in a native gibberellin pathway leading to a product necessary for normal growth. The non-responsive nature of the other five mutants to all known gibberellins suggests that there are other reasons for dwarfism than merely a gibberellin treatment. There is at present no evidence, however, for inhibitor accumulation to explain the dwarf habit of growth, at least for the gibberellin-responding dwarf mutants of *Zea mays.*

Light inhibition

Many plants exhibit inhibited growth following exposure to low intensity red radiation (540-695 mμ), and this inhibition is reversed by subsequent exposure to far-red radiation (695-800 mμ). Lockhard (1956) reported that the red light induced inhibition of epicotyl elongation in etiolated seedlings of *Pisum sativum* could be specifically reversed by gibberellin treatment. In subsequent years significance has been given to the fact that the growth of irradiated and fully etiolated plants converged to a common maximum at identical saturating dosages of gibberellin, thus the ability of the plant to respond to gibberellin is dependent on its exposure to red light. These observations have led to the interpretation that the red light-induced inhibition of stem elongation is a direct consequence of a lowered availability of gibberelline-like substances but today no determinations have been reported of the amounts of gibberellin like substance from such system. This interpretation has, however, not met universal acceptance, since gibberellin and red light appear to act independently in certain other stem growth systems. In at least one plant, *Brassica alba* (*Sinapis alba*), stem elongation inhibited by red light has not been reversed by gibberellin A_3 treatment Gibberellin and red light both promote the germination of seeds of *Lactuca sativa.*

Seed and fruit growth

Parthenocarpy, induced by gibberellin has been reported in *Lycopersicum esculentum*, *Cucumis sativus* and *Solanum melongena.* In case of *Zephyranthes* the parthenocarpic fruit contained normal appearing seeds which were devoid of embryos. The growth of ovary and integuments were specific for gibberellins. Reduced grain yield has often been reported in bakanae disease as well as with exogenous gibberellin treatments. Hayashi *et al.* (1953) found that 2 mg/i GA reduced rice grain production 32 percent, although the yield of straw increased by 14 percent. Wittwer and Bukovac (1957) found gibberellin to be approximately 500 times more effective than IAA in inducing parthenocarpy of tomatoes. It appears from several examples that native gibberellins of the seed are one of the controlling factors for seed growth. Most direct information is available on the apparent level of native gibberellins in growing seed. Corcoran (1959) has shown that increases in extractable gibberellin-like substances from the seed occur during the growth of the seed and after the virtual completion of pericarp growth.

Detached plant parts

The use of detached plant parts is often advantageous in the study of gibberellin responses. Under such conditions the investigator is able to define more precisely the environment under which the material is grown. This has, however, the disadvantage in that the limiting growth requirements must be determined before the system can be studied.

Among the detached plant parts,(i) stem sections (ii) leaf sections and leaf disks, and (iii) tissue explants and tissue cultures are used. Gibberellin affects the growth of such plant parts in various ways and these studies have the obvious advantage of being able to demonstrate and analyze the interaction of the several growth factors which may be involved.

Chlorosis

Yellowing symptoms, typical of the bakanae disease are also produced by gibberellins. Fewer chloroplasts are present in diseased tissue and the chlorophyll decreases Gibberellin treatments also decrease the percent chlorophyll. It has been shown by the I.C.I group (1954) that increase nutrients markedly reduced gibberellin chlorosis. It would be really interesting to know which of the nutrients are most important.

Metabolism

It has often been reported that when gibberellins promote elongation, they do not always cause a parallel increase in dry weight. The I.C.I group has showed that the total dry weight of wheat and peas is increased by gibberellic acid treatment, provided that sufficient nutrients are available and the experiments cover a long enough period. They also emphasized that redistribution of material from the root to the shoot must be taken into account. The increased dry weight was, however, traced to more carbon fixation and the largest carbon changes involved carbohydrates. The nature of the dry weight changes as effected by gibberellins is as yet a complex one.

The Tokyo group has indicated a decrease in total sugars, dextrin, reducing sugar, and sucrose intreated rice plants. On the other hand, the British group reported marked increase in sucrose, fructose, and glucose in wheat; in peas only glucose was affected.

No starch was found and it was suggested that protein or cellulose must increase.

Recently Hayashi et al. (1956) have examined changes of enzymatic activity with time during the development of the seventh leaf of rice. Phosphatase, alkaline, pyrophosphatase, acetylesterase, maltose, β-glucosidase, α-galactosidase amylase, urease dipeptidase, ascorbidc acid oxidase, and catalase decreased on a fresh weight basis more rapidly than in non-treated controls, but peroxidase and invertase increased more rapidly than controls and gibberellin was shown to have no effect on their action. However, these enzymatic changes did not differ in any other way from the controls except for the fact that the changes were more rapid and more marked.

The effect of gibberellin on plant respiration has been little studied. Hayashi (1940) found no effect of gibberellin on yeast fermentation . Kato (1956) recorded a 15 percent increase in oxygen consumption in growing pea stem sections, the promotion was not found in non-growing tissue. He (1957) also found that cyanide, arsenite, and p-chloromercuribenzoate inhibited gibberellin-induced pea stem elongation in low concentrations whereas copper enzyme poisons did not. He thus concluded that gibberellin did not change the terminal oxidase of the tissue.

Relationship of Gibberellin to Auxin

Although gibberellins are similar to auxins in that both promote cell elongation, it is apparent that they differ from previously known auxins in several ways. They do not in short term experiments inhibit root growth, they inhibit rather than promote root initiation, they do not cause typical epinasty nor callus formation, their action on leaf growth or dwarfism is not paralleled by auxin, and they are more potent than auxin in inducing parthenocarpy. Several other properties of auxins have been compared with gibberellin. Brian et al. (1955) could find no effect of gibberellic acid on water uptake by potato tuber disks, nor did it cause any delay of leaf abscission, also it broke potato dormancy. Perhaps the most striking of all the differences is the evident indifference to light of the gibberellin effect. So there is good reason to believe that gibberellins are not auxins in the true sense of the term.

Kinetin and Cytokinins

Up to now, we have been concerned with growth hormones, synthetic and natural, whose primary function is to promote cell elongation. We have mentioned that IAA and gibberellin do promote an increase in the number of cells in certain circumstances, but this is the exception rather than the rule. The only growth regulator we have discussed so far, concerned primarily with cell division, was traumatic acid (wound harmone). However, in addition to traumatic acid, there exists in plants several compounds capable of promoting cell division. For example, coconut milk was found to be very active as a stimulation of cell division by the van Overbeek et al. (1941).

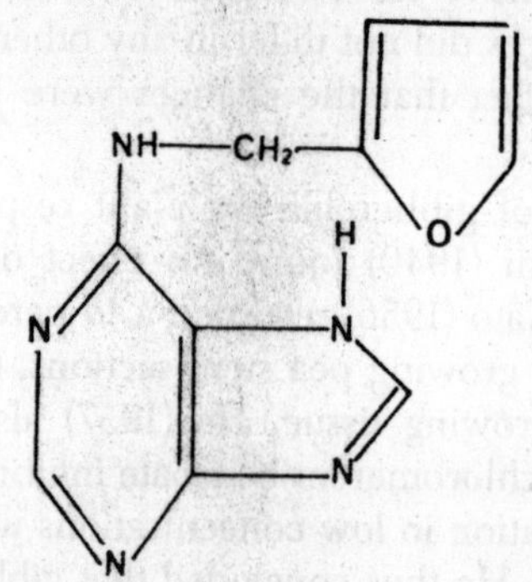

NH—CH₂ ... H ... N ... N ... N ... N ... O

Fig. 7.12. Kinetin (6-furfuryl aminopurine).

Subsequently, this finding was supported by several investigations on a variety of plant tissues, confirming that coconut milk is, indeed an active promoter of cell division. Perhaps the most exciting discovery in the search for compounds that will induce cells to multiply is *kinetin* (6- furfury-laminopurine), a compound isolated from yeast DNA by Miller et al. (1955). Actually, kinetin is formed from deoxyadenosine, a degradation product of DNA and, as such, cannot be considered natural product of plant biosynthesis. Subsequent to its discovery, many analogs of kinetin, active in promoting cell division, were synthesized. In order to group this type of substance under one title, the generic term *cytokinin* has been given to all compounds with biological activity similar to that of kinetin. As might be expected cytokinins have widerspread occurrence in the plant world.

Substances having cytokinin activity have been extracted from about 40 species of higher plants, and in most cases, the actively dividing tissues of the plants have proven to be the best sources. Evidence is also mounting for the presence of cytokinins in micro-

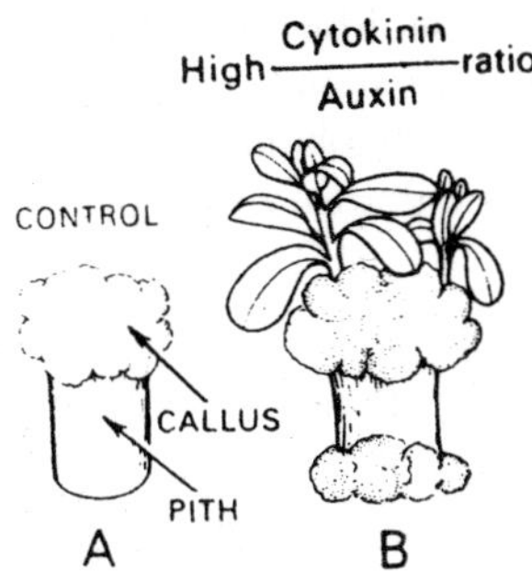

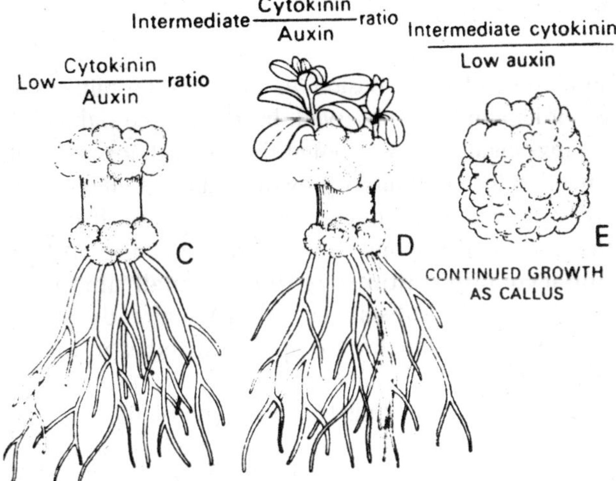

Fig. 7.13. The effect of changing ratio of cytokinin and auxin on the differentiation of buds and roots on tobacco pith cultures.

organisms. Despite the fact that many, if not all, plants contain cytokinins, it was almost 10years after discovery of kinetin before the chemical composition and properties of a natural cytokinin were described. Miller (1961) managed to extract and bring to a high state of purity a cytokinin from immature maize seeds. Attempts to crystalize and characterize the compound were unsuccessful. However, Letham (1963 : 1967) successfully extracted and purified to crystalline form a cytokinin from sweet corn. The naturally occurring cytokinin was named *zeatin.* Later, in a joint study, Letham and Miller (1965) were able to isolate in crystalline form the cytokinin that Miller had earlier isolated (in noncrystalline form) from immature maize seeds. The cytokinin proved to be zeatin. In assays for cytokinin activtty such as carrot root tissue and soybean callus cultures, zeatin proved to be much more active than kinetin.

Physiological Effects

Shortly after the discovery of kinetin, there followed a great number of papers describing its effects on many different plant growth systems. Most of these were concerned, directly, or indirectly, with kinetin's ability to pomote cell division and cell enlargement. We will discuss the effect of kinetin on cell division, cell enlargement, root initiation and growth, shoot initiation and growth, and breaking of dormancy.

Cell division

The stimulation of cell division in plant tissue cultures was the first effect of kinetin to be observed. In the tobacco pith cultures used by most investigators, it was noted that, in addition to kinetin. IAA is also needed for continuous growth. Although either growth regulator, when used one, produces a small response. It has been suggested that the small response invoked by kinetin or IAA used alone on tobacco pith cultures is due to small amounts of endogenous kinetin-like substances and IAA already present. however, when IAA and kinetin are applied together in the right ratio of concentrations, the results are striking and growth of the culture can be maintained indefinitely. This property, the stimulation of cell division, appears to be characteristics of all cytokinnis. The ability of kinetin in the presence of IAA to promote cell division.

In order that cell division may take place, an ordered sequence (DNA synthesis, mitosis, and cytokinesis) must take place. Is there a specific influence of IAA or cytokinin alone on any step in this sequence? The answer is apparently, yes Das *et al* (1956) found that both IAA and kinetin, when used alone, stimulate DNA synthesis in tobacco pith cultures. The above authors also found that both growth regulators are needed for mitosis , although IAA appears to dominate in this step. In addition, they suggested that when either kinetin or IAA is present in high concentration, the other may become limiting for at least one of the three steps needed to complete cell division. In a later paper, these authors stated that of the three steps in cell division, IAA is involved in the first two (DNA duplication and mistosis), but that the last step (cytokinesis) is controlled by kinetin. Here again, as with our discussions on gibberellins and IAA, we are presented with the importance of balance between growth hormones in plant growth and development.

How a cytokinin induces cell division is still an unsolved question. The adenine moiety of the cytokinin molecule appears essential for this process, many different substituted side chains being applicable. Strong (1958) has suggested that the side chain may influence some physical property (such as solubility) bearing on the efficiency of the growth regulator to induce cell division.

Cell enlargement

Not only do cytokinins promote cell division, but they also induce cell enlargement, an effect usually associated with IAA and gibberellin. Treatment of leaf discs cut from etiolated leaves of *Phaselous vulgaris* (bean) with kinetin causes sigificant cell enlargements. This effect of kinetin can occur in the absence of IAA. Cell enlargement after kinetin treatment has also been observed in tobacco pith cultures, tobacco roots, and in excised artichoke tissue. Stimulation of cell enlargement by cytokinins other than kinetin has also been observed. Since cytokinin-induced cell enlargement has been clearly shown, cytokinins should not be considered solely as cell division factors.

Root initiation and growth

Although relatively few studies have been made of cytokinin effects on the root system, it appears that cytokinins are able both to stimulate and to inhibit root initiation and development. Kinetin in the presence of casein hydrolysate and IAA stimulates root initiation and development in tobacco stem callus cultures. Increases in dry weight and elongation of the roots of lupin seedling were found by Frics (1960) to be promoted by kinetin. It can be seen that all concentrations of kinetin increase the dry weight of the root system even though at the higher concentration root elongation is inhibited.

In excised pea root segments, lateral root development is slightly stimulated by low concentrations of kinetin (5×10^{-5} M). At higher concentrations, however, kinetin is inhibitory. There is some evidence that interaction between cytokinins and auxin may influence the site of lateral root initiation. Bonnett and Torrey (1965), for example, demonstrated that by applying auxin and cytokinin at different concentrations to the opposite ends of excised root segments of the common bind weed (*Convolvulus*) they could alter the site of lateral root formation.

Shoot initiation and growth

In the original work with tobacco callus cultures and kinetin, it was found that callus tissue can be kept in an undifferentiated state as the proper balance of IAA and kinetin is maintained. However, if the ratio of kinetin to IAA is increased, either by the addition of more kinetin or the use of less IAA, leafy shoots are initiated. Torrey (1958) observed that kinetin initiated bud primodia on root segments of *Convolvulus arvensis* (bindweed), this effect being much more marked when the segments were grown in the dark.

Five day old bean seedlings soaked in kinetin solutions and then allowed to grow for an additional 46 hours respond with an increase in fresh weight of the epicotyl, increase in leaf expansion, and increase in the elongation of stems and petioles. In a recent study by Skoog et al (1967), auxin-cytokinin interaction in regulating growth and organ formation in tobacco callus cultres has been remarkably illustrated. The natural cytokinin, 6 (γ-γ-dimehthylallyamino) purine, is much more active than kinetin. This natural cytokinin has been found as a constituent of sRNA in yeast, corn, pea, and spinach. There have been several additional demonstrations of cytokinin induced promotions of shoot initiation and growth. However, the above studies serve to illustrate that cytokinins are active in the initiation and development of the aerial portions of the plants.

Breaking of dormancy

Earlier, we discussed apical dominance, the inhibition of lateral bud growth by auxin emanating from the apical bud. The controlling features of this phonomenon are not clearly understood and may involve not only IAA but other factors, which may interact with the auxin. This has been suggested in a study by Wickson and Thimann (1958) on the interaction of IAA and kinetin in apical dominance. They found that the growth of the lateral buds of pea stem sections in culture solutions containing IAA is inhibited, as might be expected. The growth of lateral buds on stem sections in nutrient solutions not contain in IAA, of course, is unihibited. However, the addition of kinetin along with IAA stimulates the growth of these buds. Kinetin alone has little effect. These investigators also demonstrated that the effect of kinetin on apical dominance can also be observed in entire shoots; that is, the apical bud is present. They found, as in the classical studies of apical dominance, that removal of the apical bud. stimulates the growth of the lateral buds. On the other hand, if the apical bud remains, the lateral buds are completely inhibited. However, if the instance shoot is soaked in a kinetin solution, inhibition of the lateral buds by the apical bud is overcome to a large extent. In addition to the above study, there have been several other investigations demonstrating the stimulatory influence of cytokinins on lateral bud growth. It appears that apical dominance may be controlled by a balance of concentrations between endogenous kinetin-like substance and IAA.

Other Physiological Effects

It is a well-known fact that the germination of lettuce seeds (*Lactuca sativa*) may be stimulated by red light and inhibited by infrared (far red) light. Also, the growth of bean leaf discs is sensitive to red and far red light treatment, being stimulated by red light and inhibited by far-red light treatment. In both of the above situations, the effect of kinetin treatment is similar to that of red light treatment. Only in one respect, does it differ. The stimulatory effects of kinetin are in no way inhibited by subsequent far-red treatment as in the case of red light treatment germination of seeds of white clover and carpet grass has also been observed to respond positively to kinetin treatment.

Table 7.3. Effect of kinetin and red and far-red irradiation on growth of bean leaf discs during a 48-hour growth period.

Conc.of kinetin M	*Light treatment*	*Increase in diam., mm*
0	none	1.05 ± 0.04 3
5×10^{-5}	none	2.48 ± 0.03
0	5 min red	2.48 ± 0.08
0	5 min far red	1.01 ± 0.06
0	5 min red and then 5 min far red	1.17 ± 0.07
5×10^{-5}	5 min far red	2.49 ± 0.08

Table 7.4. Effect of kinetin and red and far- red irradiation on germination of Grand Rapids lettuce seeds during a 72-hour period.

Conc of kinetin, M	*Light treatment*	*Germination %* *Expt.1*	*Expt.2*
0	none	8	7
5×10^{-5}	none	84	86
0	8 min red	96	96
0	5 min red and then		
	8 min far red	5	7
5×10^{-5}	8 min far red	86	83

Cytokinins are not only necessary factors in the growth and development of higher plants but they also markedly influence the growth of certain microorganisms. Kinetin, for example, can influence the growth of viruses, bacteria, fungi and algae all of which suggests that cytokinins are also present as natural components of lower plants. Indeed, crude extracts of natural cytokinins have been prepared from at least two micro-organisms. It is also quite evident that natural cytokinins interact with endogenous IAA to affect the growth and development of the plant. Again, we are presented with evidence supporting the concept that the growth and development of a plant is under the influence of delicate chemical balances maintained by the presence or absence of interaction between growth regulators.

Index